Anand Sharma, Sandeep Kautish, Prateek Agrawal, Vishu Madaan,
Charu Gupta, Saurav Nanda (Eds.)

Knowledge Engineering for Modern Information Systems

De Gruyter Series on Smart Computing Applications

Edited by
Prasenjit Chatterjee, Dilbagh Panchal, Dragan Pamucar,
Sarfaraz Hashemkhani Zolfani

Volume 3

Knowledge Engineering for Modern Information Systems

Methods, Models and Tools

Edited by
Anand Sharma, Sandeep Kautish, Prateek Agrawal,
Vishu Madaan, Charu Gupta, Saurav Nanda

DE GRUYTER

Editors
Dr. Anand Sharma
Mody University of Science and Technology
Lakshmangarh
Sikar-332311
Rajasthan
India
anand.glee@gmail.com

Dr. Sandeep Kautish
LBEF Campus
Opp. Maitidevi Temple
Kathmandu
Nepal
dr.skautish@gmail.com

Dr. Prateek Agrawal
School of Computer Science Engineering
Lovely Professional University
G.T. Road
Jalandhar-Delhi
Phagwara, Punjab
India
&
Institute of Information Technology
University of Klagenfurt
Universitätsstraße 65-67
9020 Klagenfurt am Wörthersee
Austria
prateek061186@gmail.com

Dr. Vishu Madaan
School of Computer Science Engineering
Lovely Professional University
G.T. Road
Jalandhar-Delhi
Phagwara, Punjab
India
vishumadaan123@gmail.com

Dr. Charu Gupta
Department of Computer Science Engineering,
Bhagwan Parshuram Institute of Technology,
GGS-Indraprastha University
Rohini, Sector 16 C
Delhi 110078
India
charu.wa1987@gmail.com

Dr. Saurav Nanda
Synopsys
690 E Middlefield Rd
Mountain View 94043, CA
USA
sauravsit@gmail.com

ISBN 978-3-11-071316-9
e-ISBN (PDF) 978-3-11-071363-3
e-ISBN (EPUB) 978-3-11-071369-5
ISSN 2700-6239

Library of Congress Control Number: 2021943689

Bibliographic information published by the Deutsche Nationalbibliothek
The Deutsche Nationalbibliothek lists this publication in the Deutsche Nationalbibliografie; detailed bibliographic data are available on the Internet at http://dnb.dnb.de.

Cover image: monsitj/iStock/Getty Images Plus
Typesetting: Integra Software Services Pvt. Ltd.
Printing and binding: CPI books GmbH, Leck

MIX
Papier aus verantwortungsvollen Quellen
FSC® C083411

www.degruyter.com

Contents

P. Megaladevi

Knowledge engineering for industrial expert systems

Abstract: Innovation is the heart of any modern manufacturing to bring high-quality, globally accepted engineering products, and automation has a vital role in industrial improvement. Knowledge engineering (KE) enables data-based application and human thought process in electronic gadgets. The present chapter deals with expert systems, knowledge-based engineering, and KE applications in modern industries by expert systems.

Keywords: Modern manufacturing, automation, knowledge engineering, expert systems

1 Introduction

Innovation is the heart of any modern manufacturing to bring high-quality, globally accepted engineering products. All manual works are automated in industries. Knowledge engineering is a field of artificial intelligence (AI) that enables applying human thought processes in machines. IT improves engineering activities; knowledge-based Engineering (KBE) is an engineering methodology and tool used to systematically work into the design system. The knowledge-based system (KBS) technology is applied to KBE in manufacturing design and production by knowledge-intensive activity through computer-aided designs (CAD) [1]. Knowledge management (KM) is applied to the entire production process. In the process, multi-disciplinary design optimization (MDO) is also used in design optimization in many disciplines. The present chapter deals with the content of applications of knowledge engineering (KE) in modern industries by expert systems. Advanced product development technologies and real-time applications are explained with examples.

1.1 Knowledge engineering

Knowledge engineering is related to expertise in human knowledge applications. Nowadays, AI is a blooming area that includes KE. KE is the technology that creates an expert

P. Megaladevi, School of Management Studies, Jai Shriram Engineering College(A Unit of Barani Hydraulics India Private Limited – An International Machine building Industry)

https://doi.org/10.1515/9783110713633-001

system to assist with issues related to their programmed field of knowledge. Applying expertise in human knowledge systems to any field involves a great transformation based on the situations. In the engineering field, with recent applications like machine learning [12, 13] and natural language processing [14] the machines learn from experience like humans. Problem-solving methods and the collateral knowledge used for each can be formed and served to solve problems, and knowledge is used to approach the issue. The amount of collateral knowledge depending on the task and the virtual expert may assist with problem identification, assisting a human, or acting as a virtual agent [2, 3].

1.2 Transfer process in knowledge engineering

Knowledge engineering transfers human expertise to the program that would absorb equivalent data and are available to an equivalent result. This is known as the transfer system. Human expertise is not always explicit, as said by the scientist. With previous experience, humans make decisions. The transfer process is left behind in the modeling process. This KE has created a new system where the traditional system, like step-by-step process, is eradicated, and they follow a clear system that provides good results in an improvement in the organization [5].

1.3 Knowledge engineering to exceed human experts

Knowledge engineering is already incorporated into the decision support software as shown in figure 1. Specialized knowledge engineers are to be employed in various

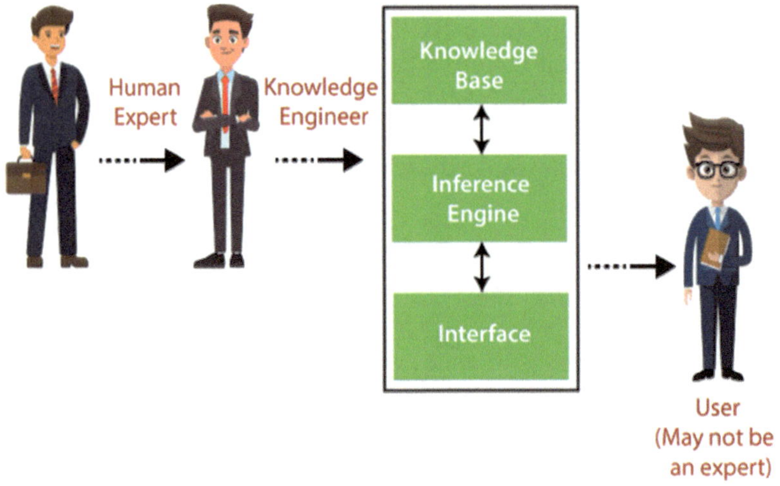

Figure 1: Knowledge engineering.

fields that are advancing human-like functions, including the power of machines to acknowledge a face or parse what an individual says for meaning. As the complexity of the model grows, the knowledge engineers might not fully understand how conclusions are being reached.

Eventually, the sector of data engineering will create systems that solve problems also as persons to at least one that does it quantitatively better than humans.

2 Knowledge-based engineering

KBE is the new technology by applying KBSs to the manufacturing domain, such as design and production. It supports CAD and is also in major activities of designing, and it is applied to the whole life cycle of the product [4, 8].

The manufacturing companies or the technologists mainly use CAD/CAM tools for NC programming. The industry is running on need-based as there are many changes in the current industry's structure and style of working. The whole process should be monitored simply by following the best system like KE, which supports industrial best practices as shown in figure 2.

Figure 2: Industry mechanization by sensors.

The need for programming automation becomes clear in everyday activities of producing engineering departments also as in workshops. By the automation of CNC machines, the programmers and operators can get significant benefits avoiding the challenges daily by solving the mentioned problems. For writing a CNC program in engineering departments, the commonly observed problems are redundant programming effort for similar sorts of parts and misinterpretation of 2D drawings and tables. The CNC programs in the manufacturing industry are needed to be perfectly matched and monitored continuously, and it has to be improvised based on the requirement. The workshop staff has to be trained to the new applications as required for the innovations. The KE system in industries can sort all these problems out. rationalizing program development, reducing the number of actions needed to finish CNC program code, and increasing program accuracy minimizes the number of inputs and selection errors [6].

Feature-based machining is an approach that allows amplifying these benefits. They allow the geometric recognition of producing features and, therefore, the automatic generation, for each of them, of the simplest (or preferred) machining process.

2.1 Scenario in industries

All manufacturing industries that deal in engineering use product lifecycle management as a tool to perform excellently in their production. Product engineering designs the merchandise from the functional and, therefore, the geometrical point of view, realizing the 3D model and, therefore, the relative drafting. Manufacturing engineers, ranging from the designed 3D model, develop and optimize machining cycles and cutting tools library and make the CNC part programs. The old/traditional CAM programming method is a slow and step-by-step process done by humans and is further verified by a team of people and again based on the requirement, needed alterations are done in the programming. However, the current KE paves the way for a very simple, easy, and accurate design. After creating the various operations, the manufacturing engineer simulates, verifies, and post-processes the tool path to get the NC-code [4, 6].

2.2 KBE methodology

KBE methodology is a method with a set of tasks to the target. It is engineering based on models. In developing KBE applications, one has to classify and implement based on the methodology. Different KBE platforms are available that support during the implementation stage through some limitations in development and maintenance. General engineering methods are used to develop the expert system for the industries. There are general software development methodologies like Rational Unified Process (RUP) or agile methods, which are used by industries. The RUP is created by the Rational Software Corporation, a division of IBM. Example: methodology and tools oriented to knowledge-based applications, EU project which proposes solutions which focus on the structuring with formalization steps as links to the implementation [1, 2].

3 Domain knowledge and representation methods

A fundamental observation from AI has been that expertise during a task domain requires substantial knowledge about the domain. A general system is to be designed to represent declarative as well as procedural knowledge. Effective inference/reasoning mechanism is essential to design an effective intelligent system. An effective or appropriate knowledge representation technique is adopted to represent the general knowledge and reasoning algorithm so that appropriate knowledge can be inferred from the system [5].

Controlled natural languages (CNLs) are an effective language used for knowledge representation. They are designed to support certain natural languages with restricted lexicon and grammar. CNLs are unambiguous and straightforward as against their base languages. Semantic web knowledge domain systems (KBSs) also contribute to deal with some broader AI problems.

Examples:

1. Explanation part (EP): Available data is separated into many parts for easy understanding.
2. Interface engine (IE): Efficient procedures and rules are used by Interface
3. Knowledge acquisition Steps (KAS): Gain knowledge about the process.
4. Knowledge acquisition (KA): Acquires information from a subject expert
5. Knowledge verification (KV): Verify and test using V- Model.
6. Knowledge representation (KR): Here, knowledge is encoded into systems with a knowledge base.
7. KBS: Knowledge base reasoning like Ontology-based reasoning is used to find solutions to problems.

In our example, the users and computer are connected to parameters with the help of the user interface. These parameters are EP, IE, KAM, KA, KVV, KR, KBS. EP, IE, and KAM are connected as a knowledge base for providing the overall structure of the system. On the other side, KA, KVV, KR, and KBS are connected with Expert system, showing the result at the knowledge engine.

4 Working memory

Working memory may be a cognitive system that is restricted with capacity that holds information temporarily. This is used as a crucial memory for reasoning and, therefore the guidance in deciding and behavior as shown in figure 3. This is called short-term memory in some cases. The data stored in the memory is sometimes manipulated, and it is short-term storage of information. Working memory stores a collection of facts

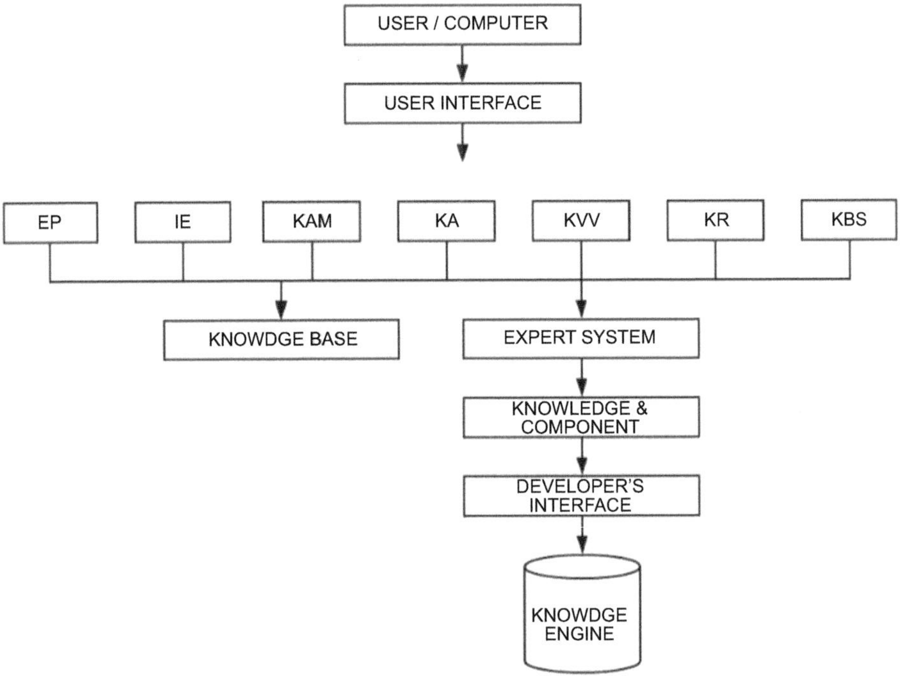

Figure 3: Flowchart of knowledge representation (etd.aau.edu.et).

- It serves as a global collection of known facts for the interface engine for a problem to give solutions one after another.

  For example, animal knowledge base

  Rule 1: Animal has hair?

  # Yes

  # Animal is a mammal

  Rule 2: Animal gives milk?

  # Yes

  # Animal is a mammal

  The reasoning process is done using iteration

5 Expert system in industry

In industries, an expert system is designed in such a way to fix complex problems by reasoning with available data, represented mainly as if-then rules instead of conventional procedural code. An expert may have special skills and knowledge in a few restricted domains or fields [9–11].

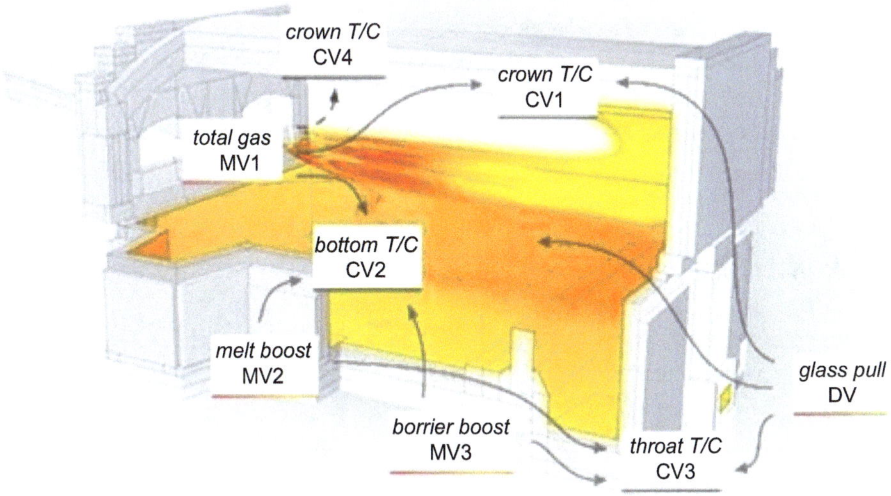

Figure 4: End fire surface temperature control glass service melting conditioning forming.

Offer a variety of advantages in comparison with human knowledge as shown in figure 4. The expert system is not perishable, but human knowledge is shifting and is influenced by various effects (e.g., their judgment are often easily impacted by new information).

5.1 Components of expert systems

Expert system is inexpensive to function, easy to replicate, and allocate documentation of the decision process. An expert system comprises

i. Knowledge consisting of domain-related facts,
ii. Knowledge consisting of domain-related rules for drawing inferences,
iii. Interpreter that applied the rules,
iv. Ordering mechanism that orders the application of rules,
v. Consistency, enforces, when new knowledge is either created or old knowledge is deleted from the knowledge base, and
vi. Justifier explains the systems by reasoning.

5.2 Characteristics and categories

Compared to standard computer applications and other AI programs, expert systems have the subsequent distinguished characteristics

- Deal with complex problems with human expertise.
- Encode human knowledge and human reasoning for a specific domain and simulatation is done by knowledge.
- Explain and validate solutions to human users at different knowledge levels.

Hence, an expert system should be ready to reason under approximate and noisy conditions.

5.3 Constructing an expert system

The phase of expert system building is time-consuming and non-trivial. The construction of an expert system consists of the following five phases:
1. Problem definition
2. Knowledge acquisition, representation, and coordination.
3. Inference mechanism
4. Implementation and
5. Learning

An expert system as shown in figure 5 is designed to learn from its experience to change and improve its current performance so that usual problems are cleared faster, and chances of new problems are reduced and solved in a better way. These can be done automatically by KE. In networking, an expert system interacts with other systems by obtaining information from a database or consulting other expert systems.

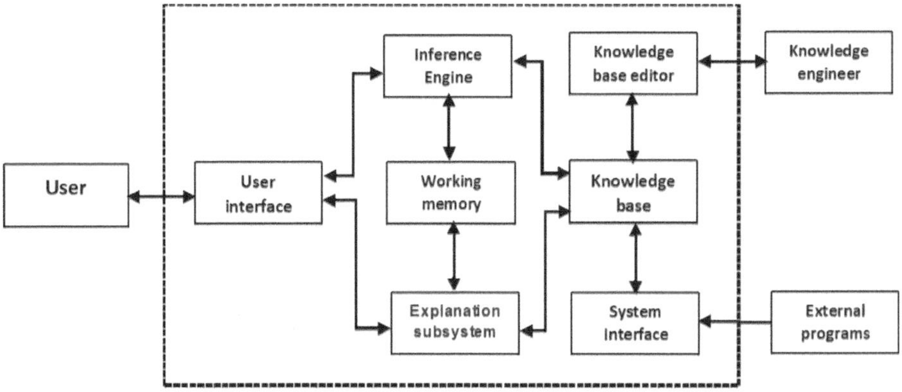

Figure 5: Architecture of a typical expert system.

Expert systems are constructed using various general-purpose programming languages also as specific tools. LISP and PROLOG have been used widely.

6 Standard Operating Sheet

It consists of all elements for a specific operation, including each step involved in the process and the approximate amount of time required for that process as shown in figure 6. It is divided into manual and robot times if the operator deals with a machine.

Manual: Work that is done by people, without the aid of machinery. Manual work is involved in loading or operating machines.

Machine: Work that is done by a machine.

Standard Operating Sheet							
S.No	Revision	Date	S/V	Operation No.	Operation Name	Area	Sht of
						Dept	Date
						Prepared by:	
S.No	Operation Description			Main Element	Key Point	Reason/ Sketch	
Safety Wear		Jigs/ Tools		Required Checks	Training Comment	Authorisation	

Figure 6: SOS model.

7 KBE and CAx

CAD is the CAx with the domain of computer-aided tools for analysis and designing. CAx is used in multiple domains like manufactured parts, software, buildings' architecture, and so on – for example, Boeing's digitally defined plane, the 777 programs, which took on the challenge of digitalization. The investment is made on large-scale systems, databases, and workstations for design and analytical engineering work.

8 Knowledge management and KBE

KM is using the knowledge of a corporation. KM is one of the most important knowledge-based technologies for KBE. KM tools support a wide range of activities that supports different types of works like large database tables, informal drawings, and notes, multimedia and hypertext objects, and so on. KM is an enabler of organizational learning which focuses on organizational objectives.

8.1 Knowledge retention

Knowledge retention is part of KM. Knowledge retention is required when expert knowledge workers leave the organization after an extended career. Retaining knowledge prevents losing intellectual capital. Knowledge retention projects are usually introduced in three stages: deciding, planning, and implementation. There are differences among researchers in terms of the stages.

9 Applications for knowledge-based engineering

There are several distinct ways to execute an automated approach or a KBE. It is important in finding a solution that best-suited application with attractive "ease of use" characteristics at the same time, such as the most durable, the simpler, and so on. Examples of various ways for automated solutions to be applied are:
- Use the KTI/ICAD IDL language to create an application.
- Make a VBA application, which many of us have done too.
- Developing an application with VB scripts.
- Using software for information ware.
- Language for PKT/GScript.
- Models for CATIA V5 [7].

10 Applications from CAD/CAE

The CAD domain has always been an early adopter of software engineering methods, such as object-orientation and rules, used in KBSs.
- Development of multi-CAD methods
- Knowledge-based building
- Simulation-driven engineering
- Alignment of systems and automation

11 Conclusion

Automation has a vital role in industrial improvement. KE enables data-based application and human thought process in electronic gadgets. The present chapter dealt with expert systems, KBE, and KE applications in modern industries by expert systems. The KBS technology is applied to KBE in manufacturing design and production by knowledge-intensive activity through CAD. KM is applied to the entire production process. In the process, MDO is also used in design optimization in many disciplines. AI and Industry 4.0 have emerged in all fields. With the application of KE, industries can develop good quality products of world-class standards.

References

[1] Baxter, D., Gao, J., Case, K., Harding, J., Young, B. et al. (2007). An engineering design knowledge reuse methodology using process modelling. Research in Engineering Design, 18(1), (doi.org), 37–48. doi: https://doi.org/10.1007/s00163-007-0028-8.

[2] Danjou, C., Le Duigou, J. & Eynard, B. (2016). Manufacturing knowledge management based on STEP-NC standard: a Closed-Loop Manufacturing approach. International Journal of Computer Integrated Manufacturing, (pdfs.semanticscholar.org) 30(9), 995–1009. doi: https://doi.org/10.1080/0951192x.2016.1268718.

[3] Devedzic, V. (2001). Knowledge modeling–State of the art. Integrated Computer-Aided Engineering, 8(3), 257–281.

[4] Guo, Y., Qasem, A. & Heflin, J. "Largescale knowledge base systems: An empirical evaluation perspective." PROCEEDINGS OF THE NATIONAL CONFERENCE ON ARTIFICIAL INTELLIGENCE. Vol.21.No. 2. Menlo Park, CA; Cambridge, MA; London; AAAI Press; MIT Press; 1999, 2006 pp 31–36.1617–1620.

[5] Harik, R. F., Derigent, W. J. & Ris, G. (2008). Computer aided process planning in aircraft manufacturing. Computer-Aided Design and Applications, 5(6), 953–962. doi: https://doi.org/10.3722/cadaps.2008.953–962.

[6] Musen, M. A. (1989). Automated Generation of Model-Basedknowledge Acquisition Tools. San Francisco, Morgan-Kaufmann.

[7] Prerau, D. S. (1990). Developing and Managing Expert Systems. Reading, MA, Addison-Wesley.

[8] Sethi, D., Agrawal, P. & Madaan, V. (2016). X-Tumour: Fuzzy Rule based Medical Expert System to Detect Tumours in Gynaecology. International Journal of Control Theory and Applications, 9(11), 5073–5084.

[9] Kaur, R., Madaan, V. & Agrawal, P. (2016). Fuzzy Expert System to Calculate the Strength / Immunity of a Human Body. Indian Journal of Science and Technology, 9(44), 1–8.

[10] Agrawal, P., Madaan, V. & Kumar, V. (2015). Fuzzy Rule Based Medical Expert System to Identify the Disorders of Eyes, ENT and Liver. International Journal of Advanced Intelligence Paradigm (IJAIP), 7(3/4), 352–367, Inderscience Publications.

[11] Kaur, P., Agrawal, P., Singh, S. K. & Jain, L., "Fuzzy Rule Based Students' Performance Analysis Expert System", International Conference on Issues and Challenges in Intelligent Computing Techniques, IEEEXplore, pp 104–109, 2014.

[12] Gupta, C., Agrawal, P., Ahuja, R., Vats, K., Pahuja, C. & Ahuja, T., "Pragmatic analysis of classification techniques based on hyperparameter tuning for sentiment analysis", International Semantic Intelligence Conference (ISIC'21), Delhi, pp. 453–459, 2021.
[13] Gupta, C., Jain, A. & Joshi, N. A Novel Approach to feature hierarchy in Aspect Based Sentiment Analysis using OWA operator. InProceedings of 2nd International Conference on Communication, Computing and Networking 2019 (pp. 661–667). Springer, Singapore.
[14] Gupta, C., Jain, A. & Joshi, N. (2018). Fuzzy logic in natural language processing–a closer view. Procedia Computer Science, Jan 1, 132, 1375–1384.

Saikat Samanta, Achyuth Sarkar, Charu Gupta, Aditi Sharma

Machine learning integrated blockchain model for Industry 4.0 smart applications

Abstract: In the last few years, machine learning (ML) and blockchain are the most prominent innovations. Blockchain's potential has been widely explored in literature and media, especially in finance and payment industries. Data confidentiality and privacy are prioritized in blockchain's decentralized database. However, this procedure is time consuming and inconvenient, which is one of the explanations why blockchain technology has yet to gain widespread acceptance. To solve the invalid dataset, we used integrated blockchain and ML approaches to secure system transactions and manage a dataset. Mostly, blockchain can greatly facilitate the exchange of training data and ML models, as well as decentralized information, stability, anonymity, and trustworthy ML decision making. We study the literature on integrating blockchain and ML systems in this paper and show how they can work together efficiently and effectively. We will go through the problems that each industry faces when it comes to implementing blockchain. We present a systematic report on ML and blockchain-based smart Industry 4.0 applications more robust to attacks in this article. Finally, we suggest some potential research avenues and anticipate further studies into the deeper convergence of the two promising technologies. We hope that our results will help decision-makers embrace blockchain technology and invest in Industry 4.0 by empowering and promoting research in this field.

Keywords: Industrial IOT; smart applications; data security and privacy; big data; cyber security.

1 Introduction

A blockchain, decentralized public ledger, stores transaction data in a series of blocks which are linked together [1]. Data has been an important source of knowledge in recent decades, bringing new possibilities to real-world challenges such as wireless

Saikat Samanta, Department of Computer Science and Engineering, National Institute Of Technology, Arunachal Pradesh, India, e-mail: s.samanta.wb@gmail.com
Achyuth Sarkar, Department of Computer Science and Engineering, National Institute Of Technology, Arunachal Pradesh, India, e-mail: achyuth@nitap.ac.in
Charu Gupta, Department of Computer Science and Engineering, Bhagwan Parshuram Institute of Technology, Delhi, India, e-mail: charu.wa1987@gmail.com
Aditi Sharma, Department of Computer Science and Engineering, School Of Technology, Quantam University, Roorkee, Uttarakhanda, India, e-mail: aditi11121986@gmail.com

https://doi.org/10.1515/9783110713633-002

networking, bioinformatics, agriculture, and finance through smart applications. Gathering as much data as possible from various sectors of the industry is one of the pillars of Industry 4.0. This produces a massive amount of data, and storing this constantly changing data in databases is a challenge, as is its connectivity, which poses security issues. BT, which has a distributed storage network, can be used to address these problems. Blockchain technology offers an immutable, trustworthy, and stable network for different organizations to share data or properties, communicate, and conduct transactions[2]. We study the literature on integrating blockchain and machine learning (ML) systems in this chapter and show how they can work together efficiently and effectively.

The German government coined the term Industry 4.0 to support the fourth wave of production by using cyber–physical technology, equipment, and processes to aid decision-making in industrial automation [3]. The modern Blockchain 3.0 development is driven by the needs of the urban world like smart cities and Industry 4.0. Forth industrial revolution has a lot of benefits and a lot of barriers to overcome in the manufacturing field.

ML is a general term that refers to a variety of approaches. ML refers to the use of machines to read, think, and act without the need for human interaction. ML can help with data authentication and detecting malicious attacks and fraudulent transactions in the blockchain.

Table 1 summarizes some recent literature analyses and studies. First, we discuss the basics of blockchain's layout and terms. We present the latest research developments in each of the associated Industry 4.0, as well as promising smart blockchain applications in these sectors. We talk about the obstacles to blockchain implementation in Industry 4.0. The paper comes to a close with some final thoughts and plans for future studies on the subject.

2 Background

In this section, we introduce blockchain technology, ML and Industry 4.0.

2.1 Blockchain

Blockchain is an immutable series of cryptographic blocks that are linked together. Hence, previous accounting ledger documents cannot be altered, and current records must be checked by a third party. Multiple transactions are received from nodes and broadcast to any node on the network in a newly generated block. Based on who has keys to the blockchain and who can check the data, the blockchain can be divided into public chains, private chains, and consortium chains [4]. Blockchain opens up new possibilities for organizing many untrustworthy actors and allowing shared governance.

Table 1: Recent literature analyses.

Year	Ref.	Technology approach	Focus
2018	[5]	Blockchain	Issues and suggestions for building blockchain-based IoT applications
2018	[6]	Blockchain and Bitcoin	Current Bitcoin confidentiality concerns and privacy-related risks to Bitcoin users, as well as an examination of current privacy-preserving solutions
2018	[7]	Blockchain and IOT	IoT encryption and privacy technologies focused on blockchain
2019	[8]	Blockchain and AI	Compile a list of new blockchain technologies, platforms, and protocols aimed specifically at the AI sector.
2019	[9]	Blockchain and Automobile	Analyzes the huge promise of blockchain technology in the automobile industry, focusing on its security capabilities.
2019	[10]	Deep learning and Machine Learning	Provide an encyclopedic overview of deep learning-based smart phone and wireless networking studies.
2019	[11]	Blockchain	Applicability of blockchain in smart cities

On the other hand, a smart contract operates on the blockchain and adds blocks. Until the blockchain network accepts a new block, it must be checked by other miners on the network using the proof of work (POW) consensus protocol. POW has the benefit of easily demonstrating that a certain amount of effort was put in to produce a block. POW entails aggregating a series of transactions into a new block and determining a hash value that is less than the target value [12].

2.2 Machine learning

To make blockchain-based apps smarter, ML's learning capability can be added. The advantages of blockchain can easily apply to ML techniques and networking communication systems [13]. Because of its decentralized nature, the blockchain has the potential to mitigate the drawbacks of centralized machine learning, such as privacy invasion, single point of failure, and scalability [14]. As seen in Figure 2, we suggested a framework for ML implementation in blockchain-based Industry 4.0 smart applications. Blockchain is used as a component in these smart applications. A blockchain network could be used to store the datasets used by ML models. In general, blockchain may be used to securely and privately archive widely trusted data base for ML systems.

2.3 Industry 4.0

Germany launched Industry 4.0, or "Industrie 4.0," in 2011 at the Hanover event. The Internet, IIOT, blockchain, big data, edge and cloud computing, robotics, human–machine interface, and artificial intelligence are all key technical for Industry 4.0 [15]. In Industry 4.0, the modernization of manufacturing processes can be accomplished by interconnected cyber-physical systems. The Internet is the most important technology in Industry 4.0 since it is the foundation for the rest of the other technology drivers. Smartness in a system is accomplished by the convergence of objects, items, and operators, as well as through the provision of context information through the Internet.

3 Use of blockchain and machine learning for Industry 4.0 smart applications

Blockchain has the potential to offer many benefits to a variety of business fields, as well as being a valuable platform for Industry 4.0 smart applications which is described in table 2.

Table 2: Blockchain's main features that can support machine learning.

Blockchain advantages	Advantages in machine learning	Detailed description
Sequential and time stamped	Sharing of data models	Data confidentiality is assured as is the freedom to audit the training phase.
Immutable, irreversible and auditable record	Updates on consensus	The training data's unchangeable records
Shared, transparent and distributed ledger	Trust for decision-making	ML's variables may be examined and audited.
Decentralized P2P network	Decentralized intelligence	In autonomous networks, machine learning may read, practice, and extract decision-making.
Cryptographic digital ledger	Confidentiality and safety	Users have complete control of how their data and trained models are shared.

Traditional databases are a fast and versatile platform for many applications on private networks. Some researchers have suggested general frameworks for deciding whether to use blockchain, there are some specific fourth industrial revolution implementation features that necessitate a more in-depth examination [16].

Although not all Industry 4.0 technologies require decentralization, some may benefit from it, particularly when a centralized system is untrustworthy. This occurs in many sectors where there is a loss of faith in financial institutions, banks, and even government entities to a certain extent [17]. The creation of public transaction logs is also motivated by trust and transparency. Many Industry 4.0 systems strictly adhere to this method, storing any completed transaction in order to conduct audits, maintain reliable traceability databases, or use big data techniques or predictive analytics [18].

In some IIOT architectures, where nodes work together to detect particular events or complete tasks, this is very common [19]. There may be other reasons to validate the use of a blockchain since there are good options offered by server farms or clouds. The most common explanation is a client's privacy conditions or a lack of confidence in the company that runs the infrastructure [20]. Since blockchain will aid in the automation of multi-company industrial processes, it is critical to describe the term "smart contract".

Smart contract terms are built on data from third-party providers that collect data from the physical world and store it in the blockchain. Since they include multiple organizations who might or may not trust each other, Industry 4.0 applications and cryptocurrencies have several popular problems. However, these organizations vary from cryptocurrencies in several ways, such as the use of power-constrained computers that must communicate with the blockchain directly or implicitly.

4 Industry 4.0 smart applications

Figure 1 depicts the application of ML techniques to blockchain-based Industry 4.0 smart applications such as manufacturing of products, Utilities and energy, Data Exchanging, Automated Customer Service and Device Personalization. By changing consumer experiences, lifestyles, and market models, ML and BT are revolutionizing current innovations.

4.1 Manufacturing of products

To make flexible schedules at specific times, ML algorithms are being used. Industries have begun to trust blockchain based procedures because of its efficiency, stability, accountability, and enforcement checks as part of the manufacturing method. With the use of flexible algorithms, product monitoring and quality management have become highly automated. It effectively detects both defective and healthy goods, even in the most fragile conditions. ML and BT systems are being used to enhance vehicle protection and capability. The company uses blockchain technology to process data more securely and quickly, giving its customers peace of mind when it comes to parking, billing, and third-party access to their vehicle.

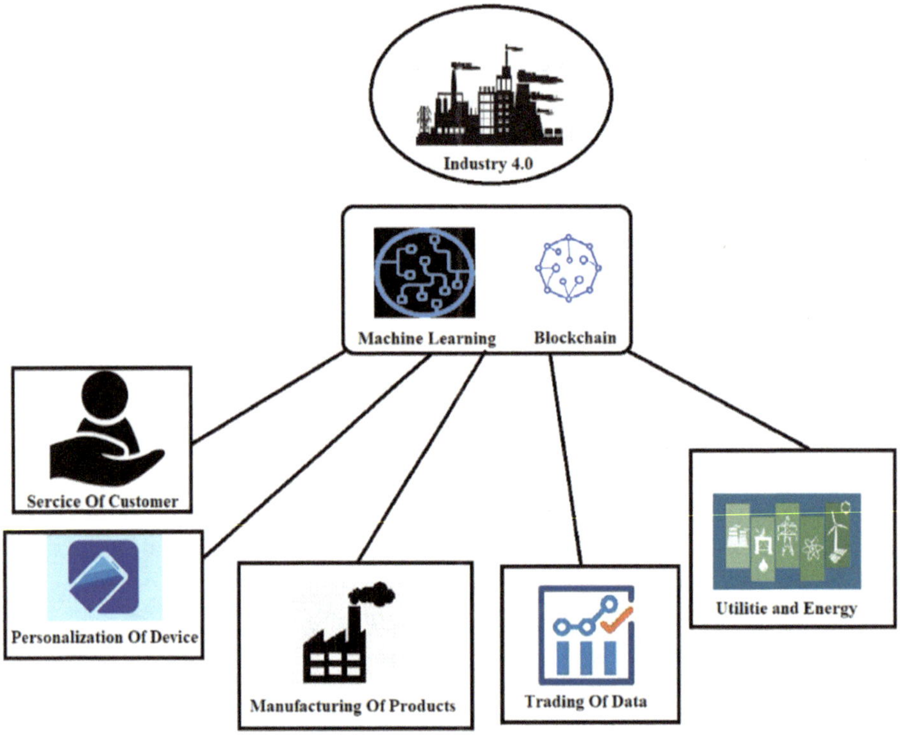

Figure 1: Industry 4.0 smart applications.

4.2 Utilities and energy

BT is helping to ease electricity exchanges in the energy sector. Smart energy renewable technologies are gaining popularity as a method of producing environmentally friendly energy products. Microgrids and smart meters, as well as smart contracts, are used to handle and register electricity transactions. The concept is to encourage businesses and individuals to exchange green energy sources without the need for a third-party intermediary. The majority of businesses are plagued by third parties that boost the cost of doing business. When the ML and BT are mixed, the revolt performs even better.

4.3 Trading of data

A data trading architecture based on ethereum has been proposed that avoids single-point loss while maintaining privacy [21]. Three groups make up the framework: a data supplier, a data user, and a business analyst. The data supplier had to make a deposit with the accountant that was higher than the amount due from the customer. The client has referred it and has asked for certain encrypted data blocks to be validated using a distance metric learning technique. There are several protocols that enable coordination between the vendor, the user, and the manager.

4.4 Personalization of devices

Smart forecasts are provided by device personalization services using device permissions. Device personalization is an element that uses predictions through smart devices to optimize quality of service for things like activities in the launcher and smart text collection when writing on the text pad. A single DP system based on ML models has been applied in a smart home environment [22]. Smart devices are attached to a smart center in this scenario.

4.5 Service to customers

With the increase in consumer demand, customer care must become reliable in order to satisfy increasing customer demands. One of the most effective methods for increasing a company's capabilities is to automate the process. For the blockchain-based application, Wang et al. [23] suggested an AutoML architecture. It is made up of six layers: organization layer, customer layer, application layer, data transfer layer, AutoML layer and blockchain layer. This system assists organizations in safeguarding their data, automating operations, and sharing data with other organizations in a mutually beneficial and secure manner.

5 The analytical framework for blockchain-integrated machine learning model

If ML algorithms improve their ability to solve complex communication and networking challenges by testing and learning, humans will find it increasingly difficult to comprehend how these systems arrive at concrete assumptions and decisions. Despite the strides made by ML in a variety of areas, a lack of confidence would seriously restrict the success of ML solutions. To solve this problem and ensure that ML solutions operate well, confidence structures and audit procedures for ML solutions should be planned well. Recently, the decentralized, secure, and cryptographic capabilities of blockchain have made it possible to document an ML's decision-making process on a blockchain while being certain that the records have not been tampered with. It offers a practical approach to auditing ML application processes and dramatically improving the trustworthiness of data and templates in order to achieve public trust [24].

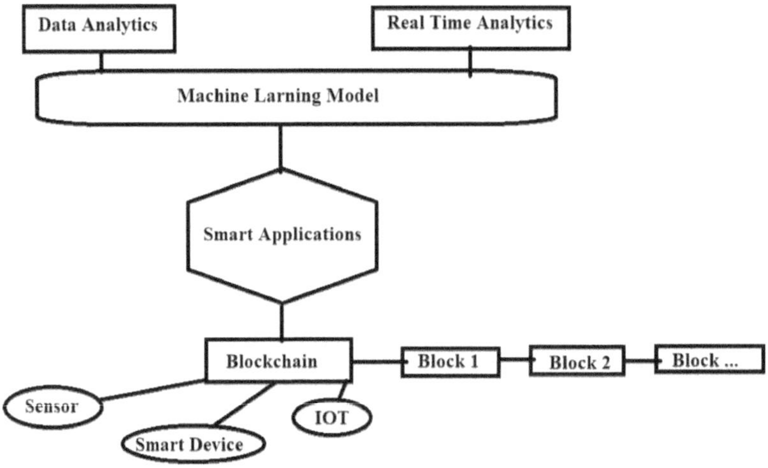

Figure 2: Frame work for blockchain integrated machine learning model.

ML techniques based on blockchain will enable the trust and safe exchange of information and decision outcomes among a large number of autonomous agents, allowing them to participate, collaborate, and vote on future decisions [25]. Transparency, provability, and explainability are among the essential facets of blockchain-based confidence protection schemes investigated by the authors in [26]. We suggest a blockchain-based ML architecture that can be trusted as seen in Figure 2.

Users can check ML tasks, trained models, rules model parameters, and other meta-data correlated with the learning process using blockchain technologies [27]. The framework has since added a validation process for delivering trusted service, which it employs to validate and check staff outputs [28–31].

6 Issues and challenges

Researchers all over the world are excited about these developments, but there are still a number of roadblocks to the integration of BT and ML. There are also unresolved problems and difficulties. As seen in Figure 3, we address the future open problems and obstacles of ML integrated blockchain technology for Industry 4.0 smart applications. Security, Privacy, Capability, Infrastructure and Memory are highlighted as challenges.

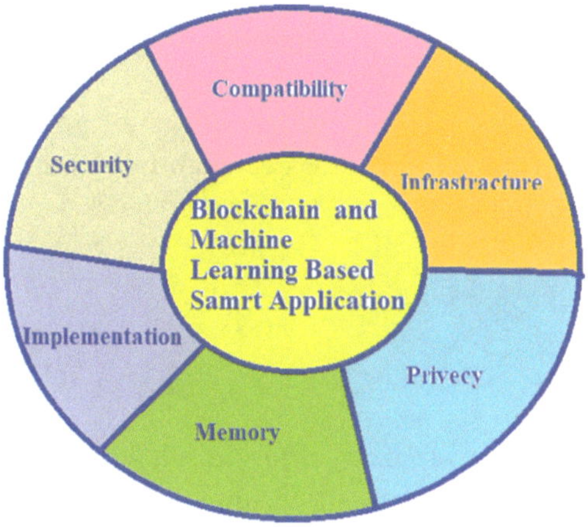

Figure 3: Issues and challenges.

6.1 Privacy

The data created by devices that will be stored on the blockchain is accessible to all blockchain nodes [29]. This raises the possibility of a privacy issue with data that has to be kept secret or confidential. Private blockchains, managed entry, and encryption may be used to solve those problems. However, the use of ML models on such small data raised challenges for forecasting and analytics.

6.2 Security

Since blockchains are decentralized, they are vulnerable to security flaws [32]. The most common fear is that the consensus protocol will be broken as a result of attacks, allowing a few farms to monitor whatever blocks are connected to the network [33]. In public blockchains, this specific issue is current.

6.3 Compatibility

If the data base cannot be trusted and the number of actors in a distributed environment is high, blockchain is a feasible alternative. A basic database is a safer choice if performance is desired. As a result, before using blockchain in any application, it is essential to understand its architecture [8].

6.4 Infrastructure

Many blockchain-based applications benefit from blockchain-specific hardware and network infrastructure. Network management, mining hardware, decentralized computing, and networking protocols are examples of these [34]. However, blockchain-specific products are also being investigated.

6.5 Memory

As new blocks are added to the blockchain, its size continues to expand. As a result, all nodes must store the whole chain, putting a significant memory limit on these machines. Irrelevant data is also a waste of computational resources. Since blockchains are permanent, data management is a major concern.

7 Conclusion

Blockchain and ML have recently advanced to the point that they are now considered ground-breaking innovations. The distributed ledger has the potential to serve as the foundation for a variety of smart applications. One of these innovations is blockchain, which has been successfully used for cryptocurrency and which, by smart contracts, will expand Industry 4.0 technologies through incorporating anonymity, confidence.

This paper looks at how blockchain and ML will work together to facilitate intelligent, stable, and ML model. We started with a quick summary of blockchain and ML, covering the fundamental principles, taxonomies, and popular implementations. Then we went through some main blockchain functionality that can help ML articles, such as data and model exchange, protection and safety, decentralized knowledge, and trustworthy decision-making. Later, we listed a number of research problems encountered during ML implementation in BT-based systems that need solutions. As a result, this article serves as a roadmap for potential Industry 4.0 technology developers to decide how blockchain and ML will improve next-generation applications.

8 Future work

Some efficiency and security problems are yet to be resolved in ML integrated blockchain technology for Industry 4.0. To begin with, the protection of BCT is determined by its implementation process and the software and hardware used in that implementation. Since all of a user's purchases on BCT are public, private information about the user can be exposed. Second, as the number of miners grows, so does the scale of the BCT. This raises the cost of data and slows delivery across the entire network, contributing to a spike in the amount of BCT scalability and capacity problems. For example, as the number of blocks is significantly expanded, the BCT's scalability becomes a problem, potentially increasing the network's latency. Despite the presence of a small number of works, a weakness in the deployment of blockchain in the 5G mobile system and subsequent implementations, such as 6G, is a major obstacle for researchers in the near future.

References

[1] Dinh, T. N. & Thai, M. T. (2018, Sep). AI and blockchain: A disruptive integration. Computer (Long. Beach. Calif), 51(9), 48–53. doi: 10.1109/MC.2018.3620971.

[2] Mermer, G. B., Zeydan, E. & Arslan, Ş. S., "An overview of blockchain technologies: Principles, opportunities and challenges," in 26th IEEE Signal Processing and Communications Applications Conference, SIU 2018, Jul. 2018, pp. 1–4, doi: 10.1109/SIU.2018.8404513.

[3] Thoben, K. D., Wiesner, S. A. & Wuest, T. (2017). 'Industrie 4.0' and smart manufacturing-a review of research issues and application examples. International Journal of Automation Technology, 11(1), Fuji Technology Press, 4–16. doi: 10.20965/ijat.2017.p0004.

[4] Zheng, X., Mukkamala, R. R., Vatrapu, R. & Ordieres-Mere, J., "Blockchain-based personal health data sharing system using cloud storage," Nov. 2018, doi: 10.1109/HealthCom.2018.8531125.

[5] Fernández-Caramés, T. M. & Fraga-Lamas, P. (2018, Institute of Electrical and Electronics Engineers Inc.). A review on the use of blockchain for the internet of things. IEEE Access, 6, May 30, 32979–33001. doi: 10.1109/ACCESS.2018.2842685.

[6] Conti, M., Sandeep, K. E., Lal, C. & Ruj, S. (2018, Oct). A survey on security and privacy issues of bitcoin. IEEE Communications Surveys and Tutorials, 20(4), 3416–3452. doi: 10.1109/COMST.2018.2842460.

[7] Ferrag, M. A., Derdour, M., Mukherjee, M., Derhab, A., Maglaras, L. & Janicke, H. (2019, Apr). Blockchain technologies for the internet of things: Research issues and challenges. IEEE Internet of Things Journal, 6(2), 2188–2204. doi: 10.1109/JIOT.2018.2882794.

[8] Salah, K., Rehman, M. H. U., Nizamuddin, N. & Al-Fuqaha, A. (2019). Blockchain for AI: Review and open research challenges. IEEE Access, 7, 10127–10149. doi: 10.1109/ACCESS.2018.2890507.

[9] Fraga-Lamas, P. & Fernández-Caramés, T. M. (2019). A review on blockchain technologies for an advanced and cyber-resilient automotive industry. IEEE Access, 7, 17578–17598. doi: 10.1109/ACCESS.2019.2895302.

[10] Zhang, C., Patras, P. & Haddadi, H. (2019). Deep learning in mobile and wireless networking: A survey. IEEE Communications Surveys and Tutorials, 21(3), 2224–2287. doi: 10.1109/COMST.2019.2904897.

[11] Xie, J. et al. (2019, Jul). A Survey of blockchain technology applied to smart cities: research issues and challenges. IEEE Communications Surveys and Tutorials, 21(3), 2794–2830. doi: 10.1109/COMST.2019.2899617.

[12] Saleh, F. (2021, Feb). Blockchain without Waste: Proof-of-Stake. The Review of Financial Studies, 34(3), 1156–1190. doi: 10.1093/rfs/hhaa075.

[13] "IEEE Xplore Full-Text PDF:" https://ieeexplore.ieee.org/stamp/stamp.jsp?arnumber=8359287 (accessed Apr. 14, 2021).

[14] Louridas, P. & Ebert, C. (2016, Sep). Machine Learning. IEEE Software, 33(5), 110–115. doi: 10.1109/MS.2016.114.
[15] Zhou, K., Liu, T. & Zhou, L., "Industry 4.0: Towards future industrial opportunities and challenges," in 2015 12th International Conference on Fuzzy Systems and Knowledge Discovery, FSKD 2015, Jan. 2016, pp. 2147–2152, doi: 10.1109/FSKD.2015.7382284.
[16] Lo, S. K., Xu, X., Chiam, Y. K. & Lu, Q., "Evaluating suitability of applying blockchain," in Proceedings of the IEEE International Conference on Engineering of Complex Computer Systems, ICECCS, Feb. 2018, vol. 2017-November, pp. 158–161, doi: 10.1109/ICECCS.2017.26.
[17] Cai, H., Xu, B., Jiang, L. & Vasilakos, A. V. (2017, Feb). IoT-Based big data storage systems in cloud computing: perspectives and challenges. IEEE Internet of Things Journal, 4(1), 75–87. doi: 10.1109/JIOT.2016.2619369.
[18] Marjani, M. et al. (2017). Big IoT data analytics: architecture, opportunities, and open research challenges. IEEE Access, 5, 5247–5261. doi: 10.1109/ACCESS.2017.2689040.
[19] Preden, J. S., Tammemae, K., Jantsch, A., Leier, M., Riid, A. & Calis, E. (2015, Jul). The benefits of self-awareness and attention in fog and mist computing. Computer (Long. Beach. Calif), 48(7), 37–45. doi: 10.1109/MC.2015.207.
[20] Barhamgi, M., Perera, C., Ghedira, C. & Benslimane, D. (2018). User-centric Privacy Engineering for the Internet of Things. IEEE Cloud Computing, 5(5), 47–57. doi: 10.1109/MCC.2018.053711666.
[21] Zhao, Y., Yu, Y., Li, Y., Han, G. & Du, X. (2019). Machine learning based privacy-preserving fair data trading in big data market. Information Sciences (Ny), 478, 449–460. doi: 10.1016/j.ins.2018.11.028.
[22] Fernández-Caramés, T. M. & Fraga-Lamas, P. (2018, MDPI AG, Dec. 01). Towards the internet-of-smart-clothing: A review on IoT wearables and garments for creating intelligent connected E-textiles. Electronics (Switzerland), 7(12). doi: 10.3390/electronics7120405.
[23] Pérez-Expósito, J. P., Fernández-Caramés, T. M., Fraga-Lamas, P. & Castedo, L. (2017, Mar). Vinesens: An eco-smart decision-support viticulture system. Sensors (Switzerland), 17(3). doi: 10.3390/s17030465.
[24] Corea, F., "Introduction to Data," 2019, pp. 1–5.
[25] Wang, S., Yuan, Y., Wang, X., Li, J., Qin, R. & Wang, F. Y., "An overview of smart contract: architecture, applications, and future trends," in IEEE Intelligent Vehicles Symposium, Proceedings, Oct. 2018, vol. 2018-June, pp. 108–113, doi: 10.1109/IVS.2018.8500488.
[26] "Beyond the Hype: On Using Blockchains in Trust Management for Authentication | IEEE Conference Publication | IEEE Xplore." https://ieeexplore.ieee.org/document/8029486 (accessed Apr. 13, 2021).
[27] Wang, T., "A unified analytical framework for trustable machine learning and automation running with blockchain," in Proceedings – 2018 IEEE International Conference on Big Data, Big Data 2018, Jan. 2019, pp. 4974–4983, doi: 10.1109/BigData.2018.8622262.
[28] Agrawal, P., Madaan, V., Roy, A., Kumari, R. & Deore, H. FOCOMO: Forecasting and monitoring the worldwide spread of COVID-19 using machine learning methods. Journal of Interdisciplinary Mathematics, 1–25. doi: 10.1080/09720502.2021.1885812.
[29] Singh, M. & Kim, S. (2018, May). Trust Bit: Reward-based intelligent vehicle commination using blockchain paper. IEEE World Forum on Internet of Things, WF-IoT 2018 – Proceedings, 2018-January, 62–67. doi: 10.1109/WF-IoT.2018.8355227.
[30] Gupta, C., Agrawal, P., Ahuja, R., Vats, K., Pahuja, C. & Ahuja, T., "Pragmatic analysis of classification techniques based on hyperparameter tuning for sentiment analysis", International Semantic Intelligence Conference (ISIC'21), Delhi, pp. 453–459, 2021.

[31] Gupta, C., Jain, A. & Joshi, N. (2019, Jan 1). "DE-ForABSA: a novel approach to forecast automobiles sales using aspect based sentiment analysis and differential evolution". International Journal of Information Retrieval Research (IJIRR), 9(1), 33–49.
[32] Kumari, A., Tanwar, S., Tyagi, S. & Kumar, N. (2019, May). Verification and validation techniques for streaming big data analytics in internet of things environment. IET Networks, 8(3), 155–163. doi: 10.1049/iet-net.2018.5187.
[33] Lin, J., Yu, W., Zhang, N., Yang, X., Zhang, H. & Zhao, W. (2017, Oct). A Survey on Internet of Things: Architecture, enabling technologies, security and privacy, and applications. IEEE Internet of Things Journal, 4(5), 1125–1142. doi: 10.1109/JIOT.2017.2683200.
[34] Chen, Z., Li, C. & Sun, W. (2020). Bitcoin price prediction using machine learning: An approach to sample dimension engineering. Journal of Computational and Applied Mathematics, 365, 112395. doi: 10.1016/j.cam.2019.112395.

Kudirat Abiola Adegoke, Akor P. Usman, Mohamed Bitagi

Prototyping the expectancy disconfirmation theory model for quality service delivery in federal university libraries in Nigeria

Abstract: The role of university libraries in the attainment of mission and vision of their parent institution can never be over emphasised. This has motivated the researchers to design a prototype for Expectancy Disconfirmation Theory (EDT) using the five dimensions of the SERVQUAL instrument to assess users' satisfaction with library services quality. The methodology used was descriptive survey, 722 postgraduate students participated in the survey while only 50 library users were tested during the usability evaluation exercise. Data was analysed with frequency counts, percentage checklist, charts using SEM, SPSS and STATA. The study discussed the Actor and use case, obtained a domain to fast track the empirical usability testing process. It was statistically proven that the usability of library electronic resources had negative relationship with users' satisfaction and service quality had a positive influence on the level of library users' satisfaction with electronic resources in federal university libraries in Nigeria and users' preferred channel of communication was text message (SMS). The study discussed the procedures involved in designing the disconfirmation software and concluded by encouraging libraries to adopt the prototype for measuring library services quality for effective service delivery. The paper recommended that libraries should improve on the quality of services delivered to their respective users through provision of effective security measures, regular evaluation of library resources and users' satisfaction, subscription to subject based resources, reviewing of library policies and regular sensitization of the users community through SMS.

Keywords: Library services, SERVQUAL, EDT Model, Users' Satisfaction

1 Introduction

Over the years, the user's service sector, including information service organizations of which university libraries are not exempted, has been vested with the responsibilities of providing effective service delivery to its ever-increasing user communities

Kudirat Abiola Adegoke, Abdullahi Fodiyo Library Complex, Usmanu Danfodiyo University, Sokoto, email: kudrat.abiola@gmail.com
Akor P. Usman Abubakar, Mohamed Bitagi, Department of Library and Information Technology, School of Information and Communication Technology (SICT), Federal University of Technology, Minna

https://doi.org/10.1515/9783110713633-003

and discharge their duties as the nerve center of academic excellence, users' satisfaction is crucial. Users' satisfaction has been reported to have influenced users' intention to use library resources, word-of-mouth behavior (recommending, advocating, and complimenting), and switching behavior (seeking for information resources elsewhere or not using library services at all). To this end, users' satisfaction assessment should be a regular exercise in the university libraries because it is possible to make resources available, but it is not possible to coerce users to use the provided resources. In other words, users are expected to be treated like kings and queens if the huge investment made on library and information resources would not amount to a waste.

University library resources and services are rendered to support their parent institutions' teaching, learning, and research activities. The university library is an integral part of an academic institution and is saddled with the responsibility of providing print and electronic information resources to support the vision and mission of the parent institution. Nigerian academic institutions, particularly the universities, were broadly categorized by the Nigerian Universities Commission [20] into three clusters in terms of ownership, namely, federal, state, and private universities. The motive behind the establishment of these universities is undoubtedly connected with the need for high-level workforce development. An essential need for human development is to seek information irrespective of the format it is presented. Therefore, university libraries have gone beyond mere storehouses of books but are now seen as the technology-driven users' information and literacy skills acquisition centers where users' information needs are turned into reality. The need for information cannot be overemphasized; in corroboration with the view of the authors, Udensi and Akor [1] posited that information is "life" – that is, information has become an essential part of everyone's daily activities because nothing happens without information.

Information provision in federal universities in Nigeria has taken a new dimension as the librarians now move from manual ways of discharging library services to digitalized methods of acquiring, processing, and disseminating library resources and services. In other words, the task of university librarians has transformed from being a caretaker of information to being an information professional who manages a system of information in multiple formats. This information is disseminated through specifically designed services as per the users' request, and the library's perception of users' information needs to justify the aim and objectives of establishing a library in any institution of higher learning. Thus, the librarians must be able to acquire relevant information resources, process, preserve, store, and disseminate these resources at the right time and to the right user using different information and communication technology (ICT) tools.

Advancement in research has brought about the application of ICT in our day-to-day activities which has also changed the library's traditional method of service delivery to a modern system of handling information right from the acquisition stage up to the dissemination and feedback stages. The libraries and information centers use computers and telecommunication devices to carry out their day-to-day operations.

Accordingly, universities and libraries are operating in a rapidly changing information environment [2]. Hence, libraries have been retooled to serve as facilitators to a wealth of information far beyond the physical collection, providing regular and instantaneous access to global information through online resources and systems. They also create their digital content for local and global access. In creating their local content, libraries subscribe to electronic databases and develop a database of Open Access resources to complement the available subscribed electronic resources to satisfy the information needs of the library users.

Considering the state of Nigeria's economy, the post-covid adverse effect is characterized by the recession, inability to fund education section, and lots more. This problem has been reported to have negatively impacted the availability of electronic resources in Nigerian federal university libraries. This has equally hampered some university libraries from subscribing to critical databases or electronic resources, making regular checking of library resources availability necessary.

This is not far from Tiemo and Ateboh's [3] report, which unveiled lack of funds as one of the factors that hindered the provision of adequate electronic resources (aggregated databases) and facilities in some university libraries in Nigeria. Similarly, the issue of inadequate funding of university libraries has made the Nigerian Library Association embark on advocacy and public sensitization programs to persuade the stakeholders to improve the university libraries' funding through the 10% of the recurrent expenditure of the universities. Ekoja [4] said that "the TETFund intervention is just a mere intervention to complement the 10% budgetary provision of the universities in Nigeria, not to replace the 10% that is supposed to be used for library development. Ekoja further explained that the TETFund is inadequate for the provision of quality library services". Meeting the information needs of users requires the provision of the right information resources and services that will satisfy their needs. If users' needs and expectations are met, then naturally, they are most likely to be satisfied. Satisfaction can be seen as the aftermath experience that the library users have regarding a favorable or positive service or product.

1.1 Statement of research problem

Because of the importance of users' satisfaction assessment in providing effective service delivery in the federal university libraries, the preliminary study has observed a communication gap between the university libraries and the user communities in the utilization of electronic resources provided by university libraries covered by the study, which has drastically affected the use of electronic library resources in the federal university libraries. It is against this backdrop that the researcher has deemed it fit to investigate the cause(s) of low utilization of the library electronic resources and services using different statistical methods to test the influence of usability and quality of library services on users' satisfaction with electronic resources as well as using the SERVQUAL questions to determine the users' expected service delivery as against the perceived service delivery.

1.2 Objective of the study

Objectives of the study are to:

i. Determine the usability of electronic resources in federal university libraries in Nigeria.
ii. Determine the quality of library services rendered by the federal university libraries in Nigeria.
iii. Design a prototype for assessing users' satisfaction with an embedded really simple syndication feeds for current awareness services using short message services (SMS).

1.2.1 Hypotheses on the influence of usability and quality of library services on users' satisfaction with electronic resources' provision in federal university libraries in Nigeria

The following null hypotheses will be tested at a 0.05 level of significance.

H_{01}: Usability will have no significant influence on library users' satisfaction with electronic resources in federal university libraries in Nigeria.

H_{02}: There is no significant relationship between library service quality and users' satisfaction with electronic resources in federal university libraries in Nigeria.

2 Review of related literature

Usability is how easy it is for library users to access, navigate, download, save, or print from the library electronic resources website or portal to satisfy their information needs? Possibly, if users get lost on the library website or database home page without any library staff to put them through, they may leave the site, or if they find it challenging to navigate from one document to another, they may feel unfulfilled. Similarly, Manandhar [5] was of the view that usability assessment encourages service providers to know their products or resources better. The inquiries such as how is the item worked by the users? Are the users judiciously utilizing the resources provided? Is the item viable to arrive at its ideal objectives? At last, is the inquiry about the item justified, despite any trouble? Therefore, the usability of electronic library resources is an essential factor toward the effective utilization of electronic resources. Availability of electronic resources in the university library does not necessitate utilization, but rather, the ease of use, accessibility, users' opinion, and relevance to users' area of study can aid the level of utilization. In corroboration, International Standard Organisation, ISO 9241-11, as quoted by Hassan [6] described usability as the extent to which specified users can use a product to achieve specified goals with effectiveness, efficiency, and satisfaction specified context of use. In a similar view, Speicher, as quoted by Adepoju, Oyefolahan, Abdullahi, and Mohammed [21], defined usability as the "effectiveness, efficiency, and satisfaction that the users achieved in using electronic information resources.

This can be evaluated through response time, ease of use, and ease of navigating the information. In corroboration with Speicher's view, Ward, Freeman and Nixon [22] said that "library collections become irrelevant if users cannot easily access them." Accessibility has to do with the extent to which library users can view, read, download, or use electronic library resources. Libraries can subscribe to electronic resources but may not know when the link to the resources will become inactive if not properly monitored and used as it is possible to see an active link today becoming inactive tomorrow due to system maintenance or upgrade.

Usability means more than just "ease of use." The five "Es"– efficient, effective, engaging, error was tolerant, and easy to learn – describe the multifaceted characteristics of usability. Interfaces are evaluated against combining these characteristics that best describe the user's requirements for success and satisfaction [7]. In corroboration, Nielsen [8]; Joo, Lin, and Lu [9]; and Matera, Rizzo, and Carughi [10] "modeled usability with five qualify constructs which are learnability, effectiveness, efficiency, memorability, error rates and satisfaction." That means that usability is about performing an excellent task within a short time using a particular electronic resource to carry out a given task successfully."

Matusiak [11] reported in a study of perceptions of usability and usefulness of digital libraries among faculty and undergraduate students of Midwestern University, United States of America. The study revealed that staff and two hundred level geography students of Midwestern University use academic full-text resources from digital libraries but later changed to open Internet sources to access visual and multimedia resources through search engines. The idea of using open Internet sources by the respondents resulted from their negative perception towards digital libraries as most of them viewed digital images. The respondents indicated that low usage of digital libraries was connected with the following perceptions: digital library systems were not user-friendly, thereby discourages them from utilizing them judiciously, academic libraries were perceived as places where textual resources are provided and used by the faculty staff and students, perception of usefulness to the respondents especially in terms of the relevance of contents, coverage, and currency, has been viewed as negative towards the use of digital libraries especially when searching for visual materials. No doubt, the review on the usability of library portals/website/ electronic resources were not judiciously utilized since most of the interfaces provided to users were not friendly."

Service quality, as described by Udensi and Akor [1] is "the standard in the library can best be determined by looking at library resources capability and utilization, meaning that the effectiveness of the library services can only be judged by its collections, facilities, and staff performance." The authors were of the "opinion that the services which satisfy a high degree information and research needs of faculty, students and other users could contribute to the success of educational and developmental goals of the institution in an effective manner." The duo further concluded that since the Nigerian Library Association is yet to design a common standard that will be used to judge the library service, especially in tertiary institutions, the quality of service delivery should be based on users' satisfaction, and if users are dissatisfied, then, the standard is perceived to below.

Lovelock and Wirtz [12] opined that "service can be viewed from 5 dimensions, namely;

1) Tangibles: refers to physical attractiveness, equipment, and materials used by the library, as well as employee performance." Tangibility in library service has to do with the conduciveness of the library spaces such as e-library space, reading areas, and collection centers. The library must be able to provide adequate facilities to enhance the teaching, learning, and research activities of the users. Library building must be located at a strategic place, possibly at the center of the university since it is regarded as the heart of the university, but when planning the library, necessary measures must be put in place to ensure that users interest is considered in whatever facility the library will provide, it must be far away from the market areas or lecture halls to avoid noise pollution and pandemonium from distracting library users and sanitation and general neatness of the environment must be ascertained. Information and communication facilities must be adequately deployed to avoid wasting user's time and for effective service delivery. Toilet facilities for normal and physically challenged people must be available and keep clean at all times. Recreation facilities such as television, satellite receivers, computer games, and constant Internet facilities must also be put in place for the retention of users. Other essential facilities should be made available to the user.
2) Reliability: This refers to the library's "ability to provide adequate service repeatedly without making any mistakes and deliver services at the right time." Users will be coming to use the library to come back to re-use the information resources; the library staff should keep a proper record of the resources consulted by users.
3) Responsiveness: concerning the willingness and ability of the employees to help users and respond to their requests, and informs when services will be provided and then provide such services quickly without wasting users' time.

4) Assurance: The behavior of the library staff that fosters customer's trust toward the library and the library can create a sense of security for the user. Assurance also means that library staff is always being polite and master every knowledge and skill needed to handle any questions or concerns from the users.
5) Empathy: It is defined as the caring, individualized attention provided to the users by the library staff. This dimension tries to explain empathy through personalized or individualized services that users are unique and special to the library. This dimension focuses on a variety of services that satisfy different needs of users, individualized or personalized services, and so on. In this case, the service providers need to know users' personal needs or wants and preferences. It implies that the library staff understands users' problems and acts in the interest of users, provides personalized attention to them, and has convenient hours of operation for maximum satisfaction.

The above five dimensions highlight what the users may likely expect as the qualities of service that can lead to effective service delivery. Thus, the five dimensions are vital to library services except for the fifth dimension that needs a minor modification for it to be appropriate for measuring library services. In the context of this study, the dimensions 1–4 will be slightly modified to reflect the variables in this study while "empathy," which is the fifth dimension, will totally be replaced with a new dimension named "effectiveness" since dimension number one (tangible) and number three (responsiveness) can take care of the "empathy" in a library setting. Therefore, it is considered unnecessary in the context of this study. The idea of effectiveness in this study is to ensure that the aim and objectives of providing library services are not being neglected. Every library tries to provide good quality services to its users, and if the services are not effectively delivered, the users will not feel the impact of a library in their teaching, learning, and research.

The dimension effectiveness will include ease of access to library resources, availability of library portal, ease of use, ease of navigating the in and outside links on library portal, regular updating of the library portal, adequate bandwidth to access the electronic resources, regular checking of the library electronic resources links, regular information literacy training for library staff and users, uninterrupted power supply, regular subscription to most used electronic resources, and adequate ICT infrastructural facilities to maximize the use of electronic resources.

The quality of resources provided by university libraries is essential to the effective utilization of the resources; an electronic resource is said to be of high quality when the credentials of the author are well known, its content covers what it is expected to cover when it is current and when it has been peer-reviewed, and when the content is not overlapping and has been accessed from a reputable or high impact database or website. All these contributed to the quality of an excellent electronic resource and will be measured by using the modified "SERVQUAL" dimensions. Users' expectations and perceptions will be measured with a 4-point Likert scale to rate their level of agreement or disagreement (1 – strongly disagree and 4 – strongly agree), in which the higher number will indicate a higher level of expectation or perceptions. Perceptions will be based on the actual service they receive in the federal university libraries in Nigeria, while expectations will also be based on past experiences and information received about library services. Service quality scores will be the differences between the users' perception and expectation scores which can be mathematically represented as SERVQUAL = P–E.

2.1 Theoretical framework

2.1.1 Expectancy disconfirmation theory

Expectancy disconfirmation theory (EDT) is a theory that has been used significantly in measuring consumer satisfaction, especially in banking and different client-related sectors of which library is not always an exemption. The principle is one of the prominent theories used for measuring how good or bad an agency is doing. The study identified the concept more appropriate for describing the difference between users' expectations and perception of the library. In corroboration with the authors' decision to use the EDT, Elkhani and Bakri [13] viewed EDT as a result of the difference in disconfirmation of expectation or users' advantageous or terrible preference, because of this that each time a person's perception of the performance of the quality of library service is higher than the expectations, it is assumed that high-quality disconfirmation has happened.

The duo pronounced that the EDT is constructed upon the idea of cognitive dissonance theory that is viewed as a dissonance between cognition of something and its truth. Similarly, while a person's notion of library service quality performance is decreased than what is predicted or preferred approximately the quality of library offerings, it indicates that negative disconfirmation has come about. Positive disconfirmation leads to users' satisfaction, and disconfirmation causes the consumer's dissatisfaction. Figure 1 shows the connection between the components of the EDT.

3 Research methodology

3.1 Research design

A mixed approach (i.e., qualitative and quantitative) will be used to describe the situation under study. The descriptive survey, observation, and interview designs will be used for this study. A mixed approach is appropriate for this study because it will investigate the influence of usability, quality of library service delivery, and users' sat-

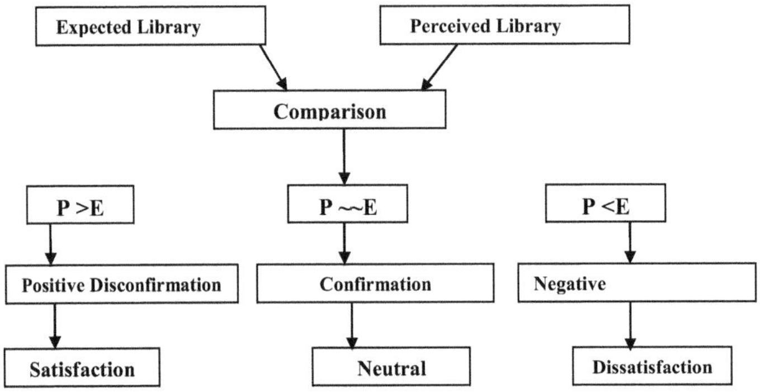

Figure 1: Expectancy disconfirmation theory model.
Source: Adapted from Oliver (1977).
Key: *P = Perception *E = Expectations > Greater than, = Equal to < Less than

isfaction by giving a clearer understanding of the variables. This study will use a focus group to conduct a direct observation of the demonstrable skills of several library users. A focus group is found suitable to observe the user's effectiveness, efficiency, and satisfaction with the library's electronic resources. The researchers see the hybrid approach as the design suitable for measuring a wide variety of unobservable data on usability and quality of service delivery and users' satisfaction and ability to collect data remotely from a large population that is too large to be observed directly.

3.2 Sample and sampling technique

The sample size for this study is 50 post-graduate users. This study applied a multistage sampling technique to arrive at the sample size that will represent the registered post-graduate users for the selected federal university libraries from the six geo-political zones of Nigeria. Purposive sampling techniques were used to select federal universities running post-graduate programs from the 43 federal universities in Nigeria. In selecting the six universities that represent the six geo-political zones of Nigeria, a stratified sampling technique was adopted, and a simple ballot system of random selection was used to select one federal university from each zone.

The following university libraries have been randomly selected to represent the entire six geo-political zones of Nigeria: University of Technology Library, Owerri, (south east), Abdullahi Fodiyo Library Complex, Usmanu Danfodiyo University, Sokoto (north west), Ibrahim Badamosi Babangida Library, Federal University of Technology, Minna (north central), Hezekiah Oluwasanmi Library, Obafemi Awolowo University, Ile-Ife (south west), Ramat Library, University of Maiduguri (north east) and John Harris Library, University of Benin, Benin City, Edo State (south-south geo-political zone) respectively.

The sample was drawn from the total population of all registered post-graduate library users for the 2017/2018 academic session.

3.3 Instruments for data collection

This study will use questionnaires and usability evaluation tools as instruments for data collection. Two sets of structured questionnaires were designed for this study; two were designed for the post-graduate students of the selected universities. An interview guide and observation checklist were used to gather more data from the participants.

3.4 Method of data collection

The researcher will embark on field trips to the six geo-political zones of Nigeria to administer the research instruments for data collection; two research assistants will be employed in each university library to help in the collection of the instruments, while the researcher will coordinate the usability testing to observe the demonstrable skills of the 10 post-graduate users of the university libraries under study. The data collection exercise will last for 8 weeks (2 months), and the data collected were analyzed immediately after the fieldwork. For usability testing, the study used the think aloud technique for evaluation of university library electronic resources (databases) in order to prepare the participants for the test; a pretest was conducted in each of the six libraries understudies to enable users to identify essential tasks and the required time for the completion of the four tasks. Interview, the study engaged the participants in a face-to-face formal interaction to further extract the dormant information from the interviewees.

3.5 Method of data analysis

Based on the design of the research instrument, the data collected will be analyzed by frequency counts, simple percentage, mean, and standard deviation using SPSS version 21, STATA, and structural equation model to determine the relationships between usability and users satisfaction.

4 Data analysis and discussions

Data were collected from two different sets of questionnaires to determine the staff and user's level of agreement with the statements usability of electronic resources in the federal university libraries in Nigeria. Findings from the data analyzed were discussed.

4.1 Availability of electronic resources in federal university libraries in Nigeria

The study found that electronic information resources such as Open Access Journals/Books, OPAC, Open Educational Resources were available in the libraries, while Oxford University Press, EduDonor, Hepseu, Oakleaf Books, Baobab, eGranary, AJOL, West Law, IEEE, Emerald, and Educational Module Contents (eLearning Platform) were not available in the libraries. Similarly, Elsevier ScienceDirect, CABDirect, JSTOR, Law Pavilion, ProQuest, EbscoHost, TEEAL, ARDI, OARE, AGORA, Lexis Nexis, Hein Online, Ebrary, Scopus, Legalpedia, IMF ELibrary, Springer, and Others: EIFL.net were not available in federal University Libraries in Nigeria. Participants were asked to log on to any of the available databases for usability tasks.

4.2 Channels of communication used by the libraries

The channels of communication are used for the dissemination of information about the availability of electronic resources to the users. These responses ranged from M = 2.14 ("University Mailing list/email alert") to M = 2.89 ("General study course-use of library"). The result shows that out of the 14 items listed for respondents to indicate how the national university library disseminates information to post-graduate students, 7 items have high mean scores above the 2.5 benchmark. These items include internal memo, Fresher's orientation, and general study course-use of the library, university website/portal, notice board, and library social platform, and university news bulletins. On the other hand, seven items produced low mean scores below the benchmark of 2.50 because they were not used to disseminate information to the post-graduate students as expected.

Consequently, those items signify that most of the respondents responded that their libraries do not use university mailing list/email alert, notice board, and electronic billboard and mobile SMS alert for disseminating information. They have never attended any sensitization workshop or training on the use of electronic resources in their libraries. Table 3 shows that most participants disagree with the questions asked about the effectiveness, efficiency, and satisfaction with the library's electronic resources and services provided.

Table 1: Method/ Channel of Information Dissemination between the Libraries and Postgraduate Users.

S/N	Items	SD	D	A	SA	N	FX	Mean	SD	Remark
		Freq. (%)	Freq. (%)	Freq. (%)	Freq. (%)					
1	University mailing list/email alert	254 (35.2)	182 (25.2)	218 (30.2)	68 (9.4)	722	1544	2.14	1.07	Disagreed
2	Internal memo	156 (21.6)	171 (23.7)	301 (41.7)	94 (13)	722	1777	2.50	0.95	Agreed
3	Sensitization workshop/training	298 (41.3)	199 (27.6)	150 (20.8)	75 (10.4)	722	1446	1.99	0.98	Disagreed
4	Fresher's orientation	134 (18.6)	151 (20.9)	332 (46)	105 (14.5)	722	1852	2.57	0.85	Agreed
5	General study course-use of library	80 (11.1)	99 (13.7)	360 (49.9)	183 (25.3)	722	2090	2.89	0.95	Agreed
6	Issuance of leaflets, brochures, pamphlets, and posters	309 (42.8)	219 (30.3)	132 (18.3)	62 (8.6)	722	1391	1.93	0.77	Disagreed
7	University website/portal	86 (11.9)	110 (15.2)	413 (57.2)	113 (15.7)	722	1997	2.77	1.14	Agreed
8	Personal interaction with librarian	154 (21.3)	187 (25.9)	330 (45.7)	51 (7.1)	722	1722	2.39	0.99	Disagreed
9	Notice board	115 (15.9)	162 (22.4)	372 (51.5)	73 (10.1)	722	1847	2.56	0.99	Agreed
10	Electronic bill board	172 (23.8)	258 (35.7)	254 (35.2)	38 (5.3)	722	1602	2.22	1.14	Disagreed
11	Database owner website	148 (20.5)	241 (33.4)	279 (38.6)	54 (7.5)	722	1683	2.33	1.25	Disagreed
12	Library social platform	118 (16.3)	234 (32.4)	291 (40.3)	79 (10.9)	722	1775	2.50	0.99	Agreed
13	Mobile alert	191 (26.5)	256 (35.5)	181 (25.1)	94 (13)	722	1622	2.25	1.00	Disagreed
14	University news bulletin	104 (14.4)	89 (12.3)	350 (48.5)	179(24.8)	722	2048	2.83	1.00	Agreed

Source: Fieldwork, 2019, Item with weighted mean value < 2.5 signifies disagree

4.3 Usability test report

Among the fifty participants for the usability testing, only ten were able to complete the three tasks given to them after pre-tasks, while the other fifty could not complete within the time frame.

Table 2: Results of the usability assessment criteria for postgraduate users.

Usability evaluation criteria	Questions related to usability factor	No. of responses	Strongly disagree A	Disagree B	Agree C	Strongly agree D
Effectiveness: simplicity, ease of access, user, friendliness, reliability	LUDUEI 1– 9	50	18 (36%)	16 (32%)	10 (20%)	6 (12%)
Efficiency: comprehensiveness, usefulness, comfort, ease of use, time saving, currency, printable	LUDUEI 1 – 7	50	14 (28%)	19 (38%)	8 (16%)	9 (18%)
			Highly dissatisfied	**Dissatisfied**	**Satisfied**	**Highly satisfied**
Satisfaction: adequacy, reliability friendliness, productivity, successful, fulfillment	LUSAQ-C4 – 2, 3, 4, 5	50	9 (18%)	23 (46%)	15 (30%)	3 (6%)
Total			**41 (27.3%)**	**58 (38.7)**	**33 (22%)**	**18 (12%)**

4.4 Interview/focus group report

A total number of seven questions were prepared to guide the interview. Data collected from the interview have been presented using a thematic approach to validate the findings gathered from the questionnaire. The table below shows the relationship between usability and users' satisfaction with electronic resources in federal university libraries in Nigeria. The result from Table 3 shows the finding of the test of the hypothesis of a relationship between usability and users' satisfaction with electronic resources in federal university libraries in Nigeria. Thus, the standardized path coefficient ($\beta = -0.028$) shows that there is a negative relationship between usability and user satisfaction, and having a p-value of ($p > 0.493$) means that the null hypothesis is accepted. Conclusively, usability had no significant influence on library users' satisfaction with electronic resources in federal university libraries in Nigeria.

5 Hypothesis Testing

H_{03}: There is no significant relationship between library service quality and users' satisfaction with electronic resources in federal university libraries in Nigeria.

Table 3 shows the relationship between usability and users' satisfaction with electronic resources in federal university libraries in Nigeria.

Table 3: Test of hypothesis one: usability will have no significant influence on the level of library users' satisfaction with electronic resources in federal university libraries in Nigeria.

Causality direction	Standardized pathcoefficient	*P* value	S.E	Decision
Usability→user's satisfaction	−0.028	0.493	0.043653	Hypothesis supported

Source: Author's calculation *using STATA; ** sig at 5%; S.E = Standard error*

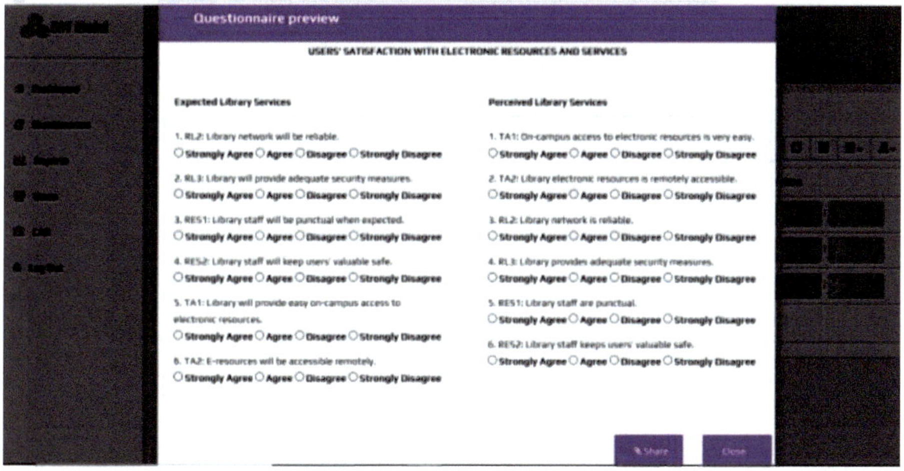

Figure 2: Questionnaire Preview Pane.

The result from Table 4 shows the finding of the test of the hypothesis of a relationship between service quality and users' satisfaction with electronic resources in federal university libraries in Nigeria. However, the standardized path coefficient ($\beta = 0.66^{**}$) depicts that there is a significant positive relationship between respondents' influence of service quality on users' research work and user's satisfaction, with a *p*-value ($p < 0.001$), which connotes the null hypothesis is rejected. Conclusively, service quality had a significant influence on library users' satisfaction with electronic resources in federal university libraries in Nigeria.

Table 4: Test of hypothesis two: there is no significant relationship between the quality of library service and users' satisfaction with electronic resources in federal university libraries in Nigeria.

Causality direction	Standardized pathcoefficient	P. value	S.E	Decision
Service quality→ user's satisfaction	0.66	0.000**	0.025313	Not supported

Source: Author's calculation using STATA; ** sig at 5%; S.E = Standard error

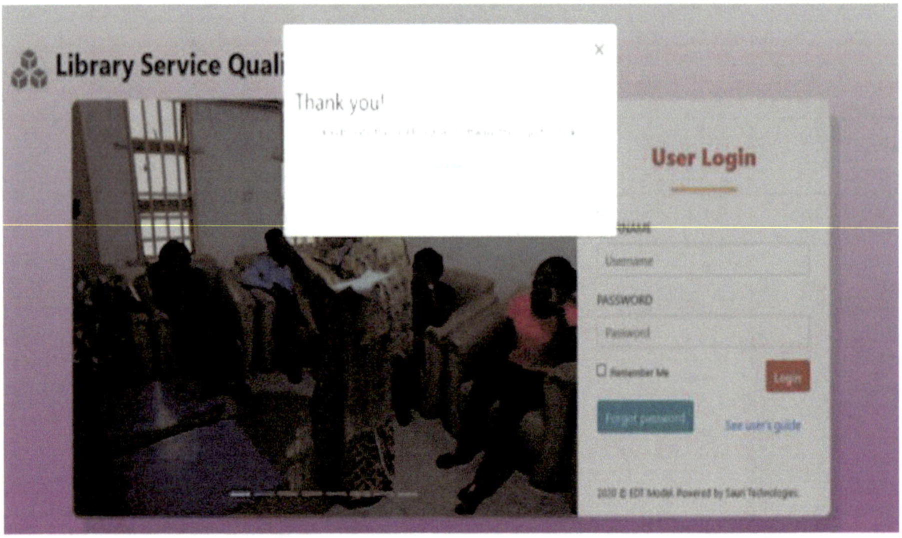

Figure 3: Feedback/dialogue box.

5.1 Expectancy disconfirmation theory process flow

1. Administrator logs into the system.
2. The administrator shares a link to the online data capturing instrument (questionnaire) using google form.
3. User visits the link.
4. User fills and submits the form*.
5. The system processes the form.
6. The system stores the data.
7. The administrator requests a report summary, search filters (i.e., gender, age range, date, session, or semester).
8. The system searches the database.
9. The system processes the data by comparing the two mean scores (one is the expected library service and the other is perceived library service).

Table 5: Thematic analysis of focus group interview with postgraduate users.

S/No.	THEME: Use of university library	THEME: Usability of electronic resources	THEME: Satisfaction
1	**Sub-theme: frequency of use of university libraries** – Just visiting the library for the first time. – Already using the library facilities. – Stopped using the library because of noise. – Library rules are too strict – Library does not have enough space for readers. – Love coming to the library to read my books. – Use library frequently for my personal reading. – Come to do my assignment in the library. – Come to library once in a while to socialize. – Really do not like coming to the library because of my valuables. – Library security measure is weak. – I use library for 4–5 days a week including Saturdays. – I do use the resources. – This is my first time in the library. – I only know about eLibrary space and that is what I am using now.	**1. Sub-theme: availability of electronic resources** – Library subscribed to relevant databases. – Library has no subscription to any electronic database. – Library fee-based e-resources are not meeting my desired information. – Library fee-based e-resources are not detailed enough. – Library fee-based e-resources are not adequate enough to answer my search queries. – Library has numerous electronic resources and few paid databases. – Library does not give me access to ebooks on CDs because they are plug and play.	**1. Sub-theme: library electronic services** – I am partially satisfied with the library e-resources because most of time I use google to get my resources. – I am satisfied with the library facilities not electronic resources. – I am satisfied with my professional database. – I am partially satisfied with so many observations. – I can say I am satisfied with so many complaints. – I am not satisfied with library electronic services at all. – Undecided because I have some complaints. – Library service is rated fairly good due to insufficient library resources. – Library electronic services have not been effective.
2.	– I use the eLibrary light to read.		– I am dissatisfied with the time allocation for the use of library café systems.

(continued)

Table 5 (continued)

S/No.	**THEME:** **Use of university library**	**THEME:** **Usability of electronic resources**	**THEME:** **Satisfaction**
	Sub-theme: **library service quality** – Library staffs are good. – Some staffs are lazy. – Some staffs are not skilled. – Sometimes library close because of lack of light. – Library does not provide alternative power supply. – Library service quality is becoming ineffective. – Library does not do selective dissemination of information through SMS alert and it's important. – Library need to do more in terms of training and retraining of users and staff. – Library need to upgrade their systems. – Stopped using the library because of lack of services in postgraduate reading room. – Many library resources are out of date. – Do not know how to search for information in the library. **Sub-theme: frequently used electronic resources.** – I use library OPAC to search for library resource. – Electronic resources are not included in the Library OPAC. – I use Open Access electronic resources for my research works. – I use Open Access resources frequently. – I use Open Electronic Resources to get information for my research work.	**2.Sub-theme: awareness** – No, I am not aware of the electronic resources. – I am not aware of any electronic resource in the library. – I am not aware. – I am aware of the electronic resources. – Library does not sensitize users. – Library informed me about e-resources. – Library sends memo to my faculty. – Library did not inform me about their available electronic resources. – Library did not send SMS to me. – Library did not organize orientation for postgraduate users. – Library should organize more sensitization program. – Majority of users are not aware of the library electronic services. – I am aware of the resources but do not know how to access it. **Sub-theme: access to library electronic resources.** – Library electronic resources can be accessed anywhere.	– I am dissatisfied with the use of e-Library due to restriction on the use of laptops and other mobile devices. – Library electronic resources cannot be effectively utilized unless there are adequate facilities for access. – Library users do not have enough time to access the electronic library resources. – I used the electronic databases once and I did not get what I wanted. – Yes, I do use the electronic resources. **2. Sub-theme:** **General library service** – Satisfied because library provides basic facilities. – Somewhat satisfied because library has conducive space for users. – Not satisfied with security measures.. – Dissatisfied with library space. – Dissatisfied with power supply in the library. – Dissatisfied with toilet facilities. – Dissatisfied with rules and regulations guiding the library. – Dissatisfied with awareness creation. – Dissatisfied with the quantity of ICT facilities in the library.

– Open Access resources contain more detailed information than ProQuest. – I use google search engine to get my electronic resources. – Google search engine is more detailed than the Ebscohost. – I use HINARI every time. – I use ScienceDirect often. – Google is better than subscribed databases. – Open sources give detailed information than the fee-based electronic resources. **Sub-theme:** **preferred channels of communication** – I prefer text message. – I prefer SMS alert. – I want Text Messages. – I like e-mail messages. – I preferred selective dissemination of information through email or SMS. – SMS is okay for me. – Text message will make me access electronic resources anywhere I am. – SMS will keep me posted. – I think SMS is suitable. – Library should use mobile text message. – Mobile messages are good particularly, SMS. – Social media is good for me, library should use WhatsApp. – Facebook will help in sharing with friends. – Library should send SMS to me, I love it.	– Library electronic resources can only be accessed within the library. – Library connection is fluctuating. Many users line up to use the library café and before it reaches one the light is already off. – Some links to electronic resources are not active. – Some databases regulate the way we download. – Library electronic resources have many overlapping title from different publishers. – Always using my data to download in the library. – Downloading of free e-resources is not free to me because I use my data. – Library disallowed Laptop in the library and their computer systems are not adequate for users of the library.	**Sub-theme: determinants of users satisfaction** – I need quiet library environment. – I want good connectivity. – I want extension of library café opening hours. – I want adequate systems with good connection. – I need specialized databases and good connection. – I need remote access to library electronic resources. – I like strong connection. – I will like library to provide adequate security measure. – I want Open Access resources to be on the desktops. – I want library to permit Laptop use in the library. – I want library to create more awareness. – I need conducive environment – Create more awareness. – I need discussion room. – Library should improve security measure. – Library should provide more computers. – Library should extend the operation hours. – I need literacy training. – Library should increase hours allocation for every user.

10. If perceived service is **greater than** expected library service, then the decision is positive disconfirmation (i.e., the user is satisfied, meaning **satisfaction** has occurred).
11. If perceived service is **equal or equivalent** to expected library service, it means **confirmation has occurred** that is **neutral (neither satisfied nor dissatisfied,** that can also be referred to as **the zone of tolerance.**
12. If **perception** is **less than expectation**, it means **negative disconfirmation (dissatisfaction).**
13. The administrator saves or prints the results.
14. The administrator navigates to share the results.
15. Administrator inputs contacts of receivers.
16. The administrator can share the results with other staff.

Flow chart of the expectancy disconfirmation model

The flow chart shows the sequence of tasks performed by the EDT model by comparing the scores of perceived library services and the users' library service expectations to determine their level of satisfaction with the library service provision as per table 5.

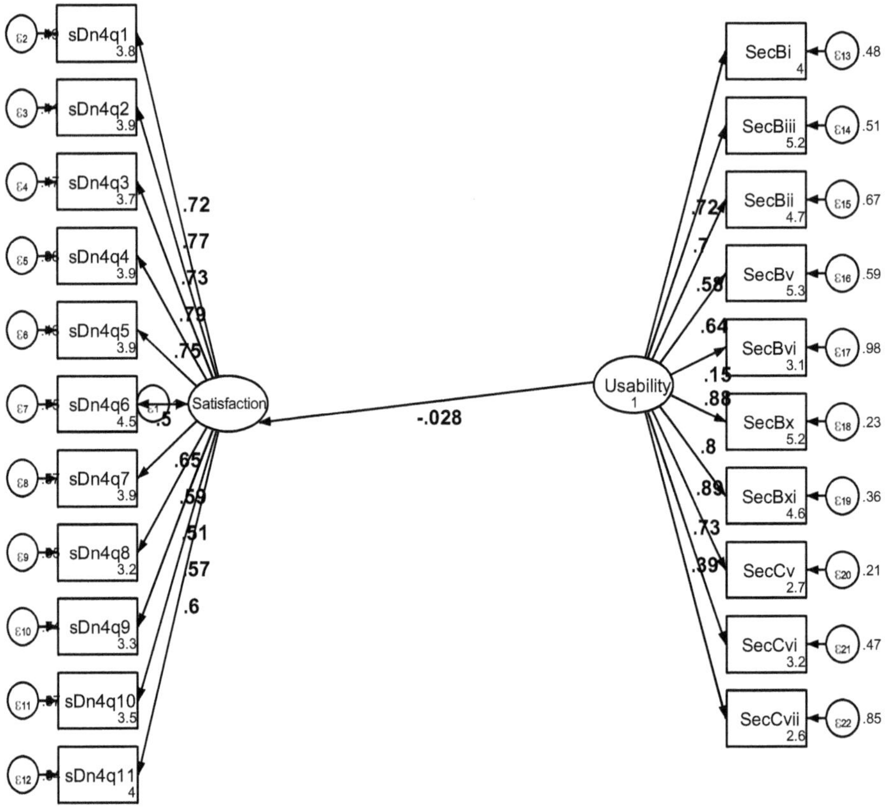

Figure 4: Structural equation model of the relationship between user's satisfaction and usability.

5.2 Library service quality assessment prototype

This prototype is designed to measure library service quality through an adapted SERVQUAL instrument (25 survey questions) – the questions were formed based on the peculiarities of the library operations and services. The five SERVQUAL dimensions are tangible, reliability, responsive, assurance and effectiveness instrument, and EDT as shown in table 6.

Table 6: SERVQUAL instruments used for developing EDT model.

1. Perceived library service quality					
S/N	**Statement**	**Strongly disagree**	**Disagree**	**Agree**	**Strongly agree**
a.	**Tangible/accessibility**				
1	On-campus access to electronic resources is very easy.				
2	Library electronic resources are remotely accessible.				
3	Library's portal/website requires less effort to use.				
4	Library environment is conducive.				
b.	**Reliability**				
5	Library sends mobile alert to users.				
6	Library network is reliable.				
7	Library provides adequate security measures.				
8	Library does not close before official time.				
c.	**Responsiveness**				
9	Library staffs are punctual when expected.				
10	Library staff keeps users' valuable safe.				
11	Library staff provides professional guidance.				
12	Library operating hours during weekend is not favorable.				
13	Library provides wheel chair services to physically challenged students.				
d.	**Assurance**				
14	Library provides literacy-training program when they promised to do so.				
15	Library provides necessary support meant for people with disability.				
16	Library subscribes to electronic resources when they promised to do so.				
17	Library has sympathy for defaulters.				
e.	**Effectiveness**				
18	Library provides adequate ICT facilities.				
19	Library updates their electronic resources regularly.				
20	Library provides adequate number of e-journal titles.				
21	Library provides adequate number of ebooks.				
22	The waiting time for library café system is suitably short.				

Table 6 (continued)

1. Perceived library service quality					
S/N	**Statement**	**Strongly disagree**	**Disagree**	**Agree**	**Strongly agree**
23	Library provides adequate bandwidth to access electronic resources.				
24	Library provides adequate computer systems for users to access electronic resources.				
25	Library provides adequate user trainings on electronic resources.				

2. Expectations: Kindly, indicate your extent of agreement with the following statements. Put a cross **(X)** on your choice of answer.

S/N	**Statement**	**Strongly disagree**	**Disagree**	**Agree**	**Strongly agree**
a.	**Tangible/accessibility**				
1	Library will provide easy on-campus access to electronic resources.				
2	E-resources will be accessible remotely.				
3	Using the library's portal/website will require less effort to use.				
4	Library environment will be conducive.				
b.	**Reliability**				
5	Library will send mobile alert to users.				
6	Library network will be reliable.				
7	Library will provide adequate security measures.				
8	Library will not close before time.				
c.	**Responsiveness**				
9	Library staff will be punctual when expected.				
10	Library staff will keep users' valuable safe.				
11	Library staff will provide professional guidance.				
12	Library operating hours during weekend will be favorable.				
13	Library will have wheel chair services for physically challenged users.				
d.	**Assurance**				
14	Library will provide literacy-training program when they promised to do so.				
15	Library will provide necessary supports meant for people with disability.				
16	Library will subscribe to electronic resources when they promised to do so.				
17	Library staff will have sympathy for defaulters.				
e.	**Effectiveness**				
18	Library will provide adequate ICT facilities.				
19	Library will update their electronic resources regularly.				

Table 6 (continued)

20	Library will provide adequate number of e-journal titles.
21	Library will provide adequate ebooks.
22	The waiting time for library café system will be suitably short.
23	Library will provide adequate bandwidth to access electronic resources.
24	Library will provide adequate computer systems for users to access electronic resources.
25	Library will provide adequate user trainings on electronic resources.

5.3 Designing the prototype

In the process of developing, this prototype was used to come up with the features and functionalities to allow for more updates and upgrades to be made in order to add more value and make an efficient tool for data collection based on the two defined services peculiarities (the expected and perceived services).

5.4 How the prototype works

This prototype provides features and functionalities with which a user can be created to access the prototype. This user can log in and perform series of tasks. These include creating more users within his/her domain. Such users created are also administrators. An administrator can create streams of questions based on their service categories (expected and perceived services). With these questions created, there is a provision for creating a questionnaire title after which questions can be picked from the pre-defined questions and finally, a questionnaire generated. Users, questions, and questionnaires can be viewed and managed. On creating a questionnaire, an option has been made for sharing the questionnaire link via email or text message. Users then follow this link and fill in their demographic information and submit.

5.5 Users process

This prototype recognizes two significant types of user: administrative user and public user. However, they are called the actors of the system, and the tasks they perform authoritatively are referred to as the user cases.

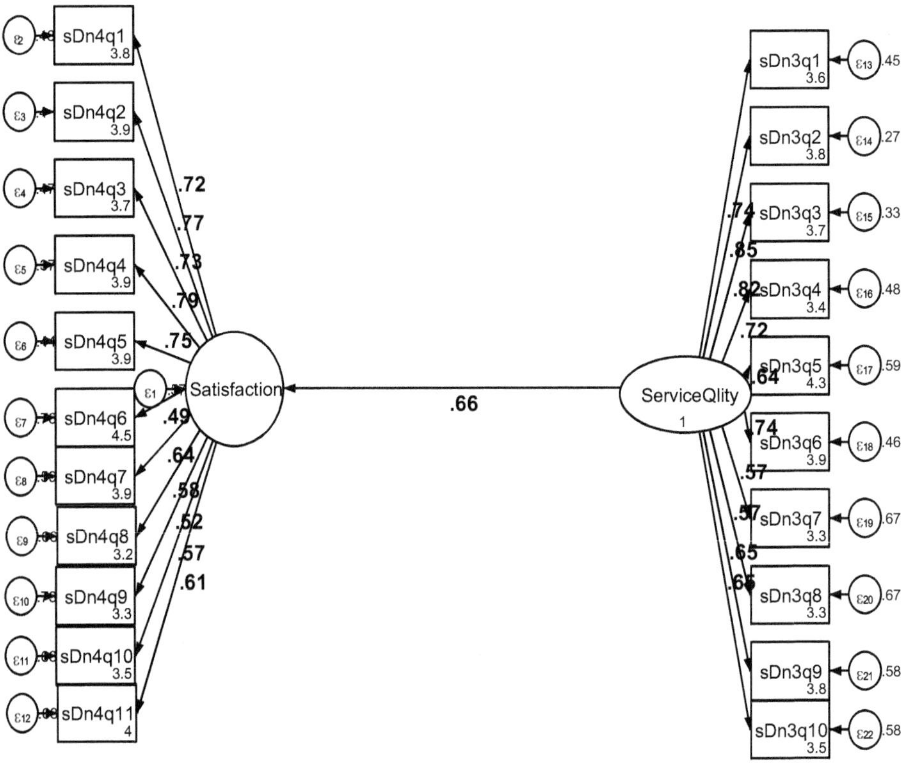

Figure 5: Structural equation model of the relationship between user's satisfaction and service quality.

5.5.1 Administrator

An authorized user who creates public users and shares the questionnaire with users. This can be a librarian, an administrative officer, or a lecturer who wants to evaluate his/her course(s).

1. The administrator logs in to the system with a pre-defined username and password by the super administrator.
   - Administrator will land on dashboard and menu options and their sub-menus as follows;

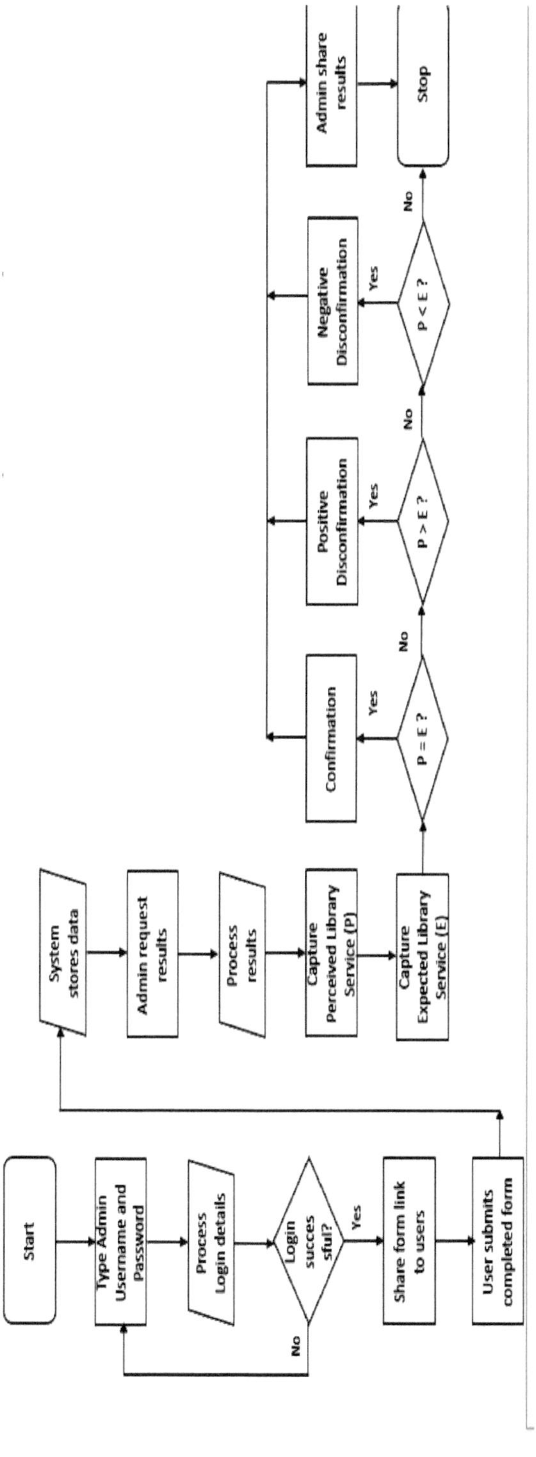

Figure 6: Flow chart of the expectancy disconfirmation model.

5.5.2 Dashboard

This provides a summary report at a glance as per the system activities. Information on the dashboard includes a total number of questions, questionnaires, administrative users, and a list of respondents.

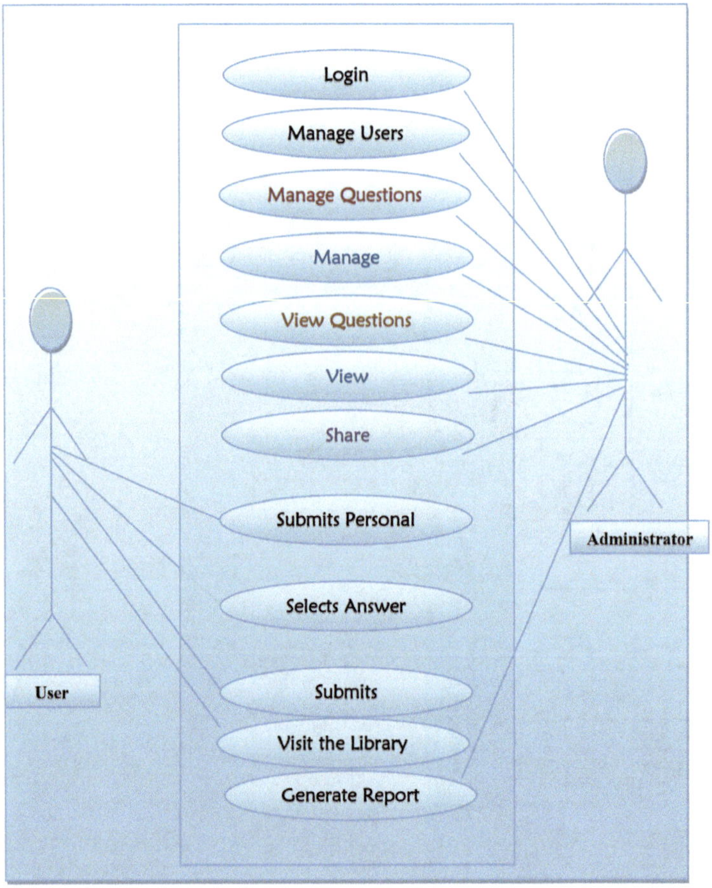

Figure 7: Actors and use case.

5.5.3 User

This is a menu that enables the admin to manage users.

Create user view: Used for creating and view other administrators who are users on the system.

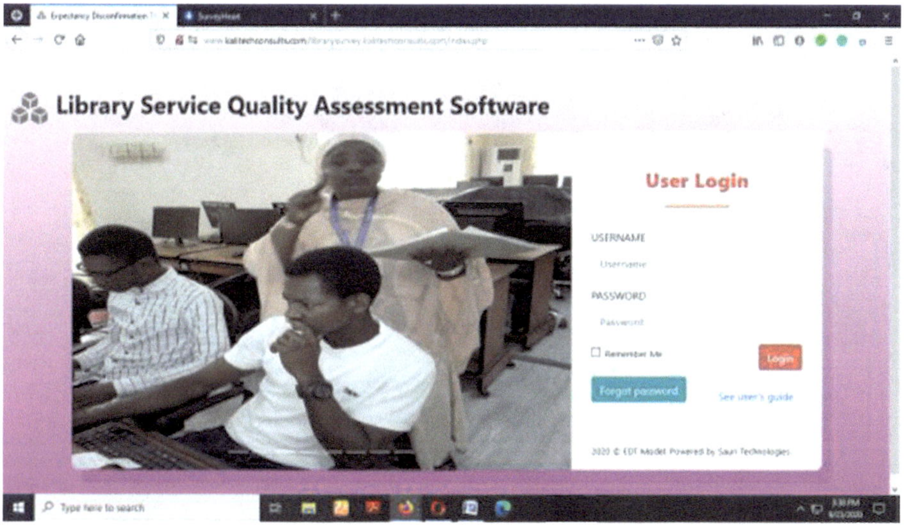

Figure 8: Log in page.

5.5.4 Process of creating user

1. Administrator logs in to the prototype
2. Admin goes to "users" main menu and goes to "create a user."
3. The administrator fills the form with the new user information and clicks the process

5.5.5 User information

– Full name, gender, email address, role (librarian, lecturer, or sales manager)

5.5.6 View user

Displays the list of available users on the system. The administrator can update or delete users' records as well.

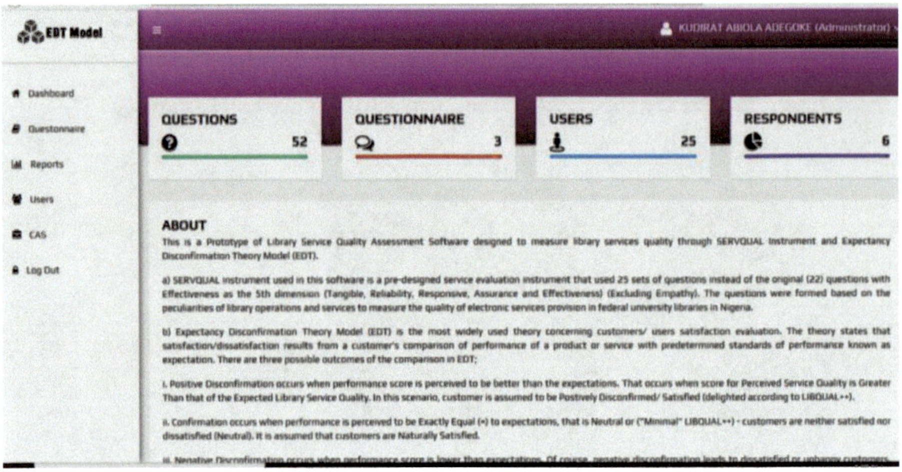

Figure 9: Dashboard.

5.5.7 Questionnaire view

This is where questions and questionnaires are created, listed, and shared

– **Create question:** This allows the administrator to create questions.

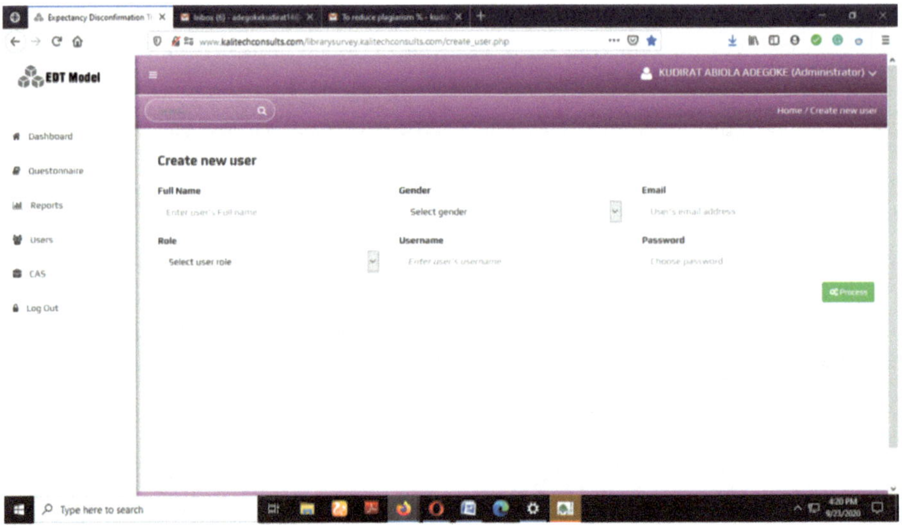

Figure 10: User account creation platform.

5.5.8 View questions

Lists all questions created by the administrator. This page also permits the administrator to edit or delete the question.

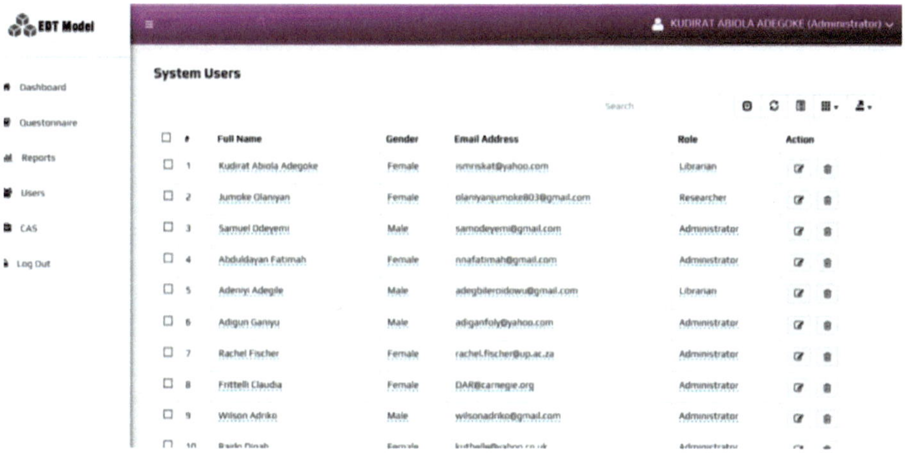

Figure 11: User link view: user information display page.

5.5.9 Create questionnaire

Allows the creation of a new questionnaire.

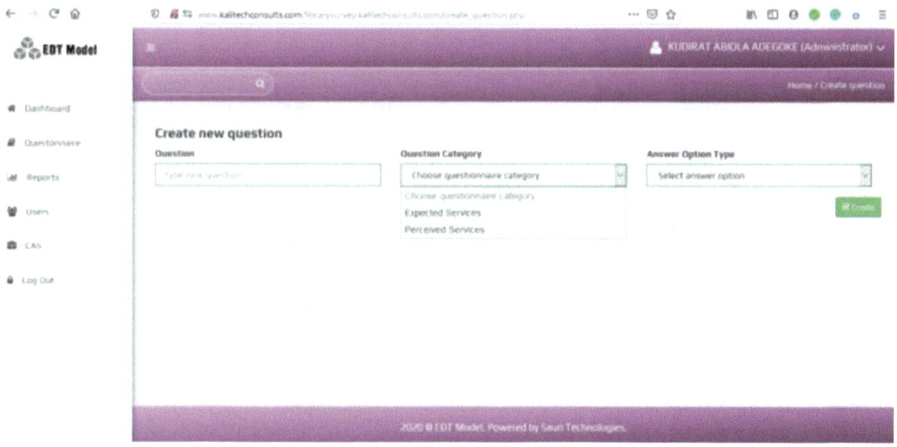

Figure 12: Question creation platform.

5.5.10 List questionnaire

Lists all existing questionnaires created by the administrator. Through this menu, a questionnaire record can be updated or deleted.

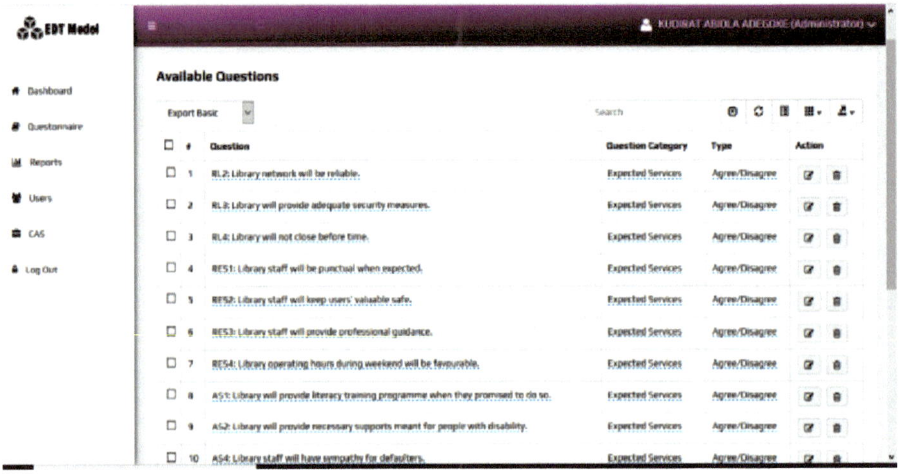

Figure 13: View questions display page.

5.5.11 Share questionnaire

Enables administrators to share questionnaire links to public users through email or text messages.

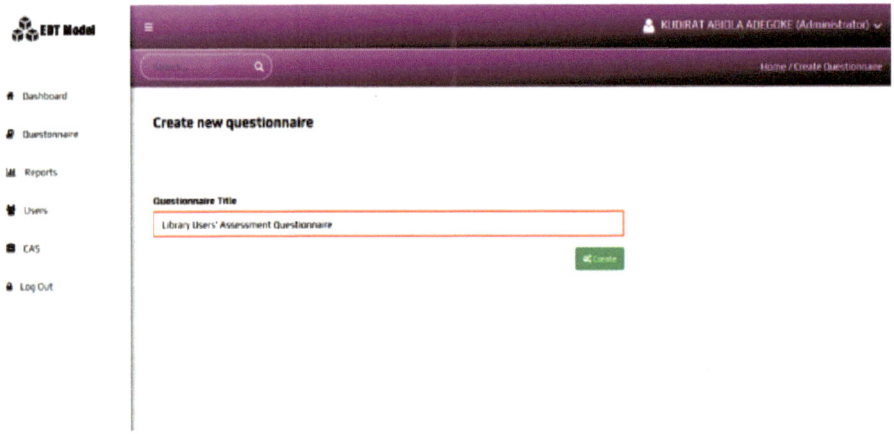

Figure 14: Create questionnaire display platform.

5.5.12 Link sharing preview

This administrator can send voice, video, or text messages to all registered users using the text.

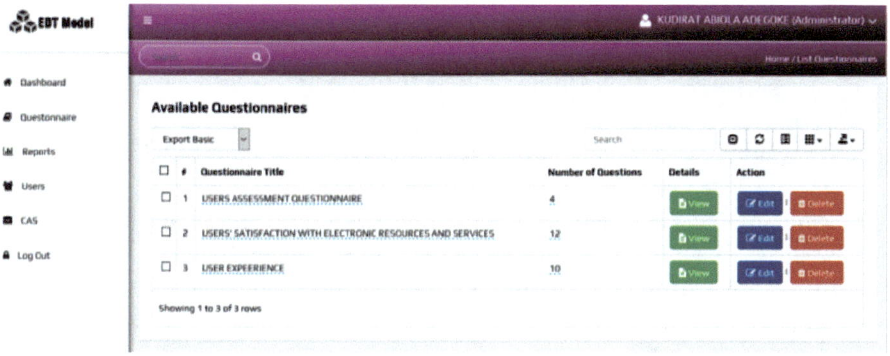

Figure 15: Questionnaire management page.

5.5.13 Report

There is a provision for an administrator to generate a report from the user responses to gather the scores obtained for a particular questionnaire shared on a particular date. To generate this report, the admin needs to select the questionnaire title, the date it was shared, the service category, and users' gender. This prototype provides a comprehensive report in tabular form and allows for the results to be exported in CSV or MS-Excel formats.

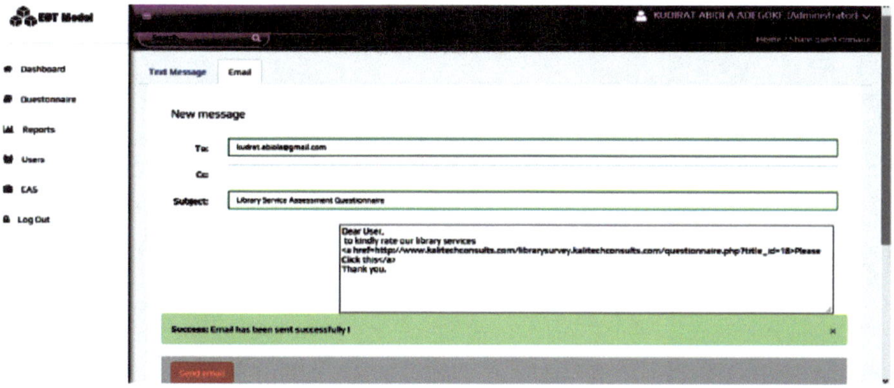

Figure 16: Questionnaire sharing page.

5.6 Findings of the study

Based on the data collected and analyzed, the study found out that:

- Majority of the participants do not use electronic library resources as expected.
- All participants preferred test message as the best channel of communication.
- Participants were not happy with the library management software's performance during the usability test.
- Many participants got frustrated and could not complete the three tasked assigned to them due to slow Internet connectivity.
- The usability of electronic library resources has no significant relationship with users' satisfaction.
- Service quality has a significant influence on library users' satisfaction with electronic resources in federal university libraries in Nigeria.
- There is no software for measuring library service quality that is embedded with SMS and therefore recommended the prototype for constant evaluation of library services.

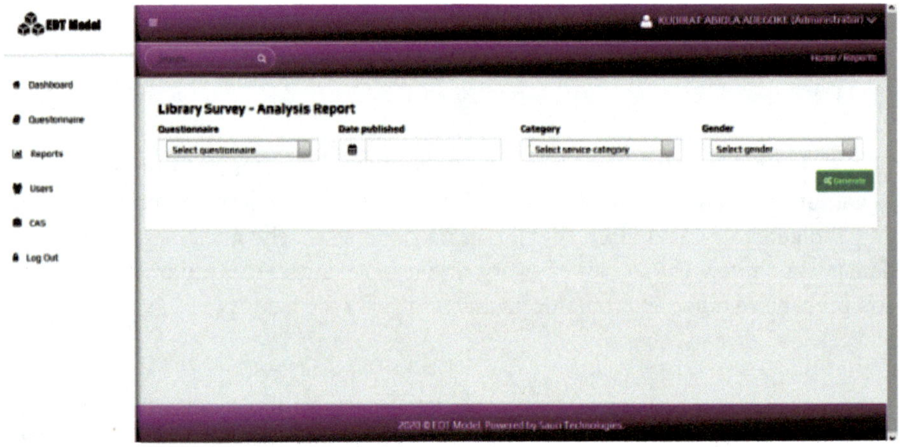

Figure 17: Report management page.

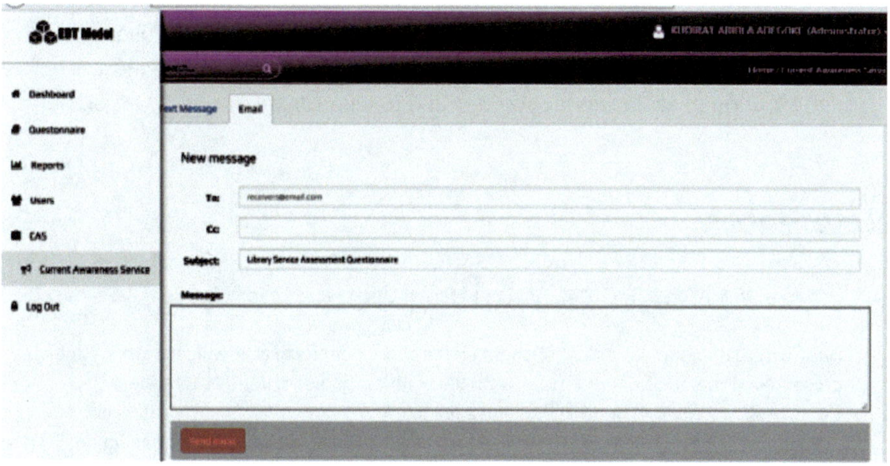

Figure 18: Current awareness services view.

6 Conclusion

Given the review of related literature and the study's findings, it is pretty clear that previous studies have failed to identify the factors responsible for the low utilization of electronic resources in federal university libraries in Nigeria. This study has found that the low utilization of the library electronic resources are connected to the unavailability of relevant electronic information resources for users' research works, other factors such as lack of remote access, ineffective communication/information dissemination of information, and lack of constant training, upskill, and retooling. The retraining of both staff and users of the library had contributed to users' dissatisfaction over the years; these have been tagged as lousy library service delivery which no doubt have been responsible for low utilization of the electronic resources in Federal university libraries in Nigeria. Thus, the study has recommended the following measures to increase users' satisfaction with library service delivery.

Recommendations

- Libraries should use SMS as the default channel for disseminating information to users.
- The library should increase its bandwidth to facilitate easy access to electronic resources. The library should advise all database owners to make the home page friendly to users.
- The library should adopt EDT software for measuring library service quality and user satisfaction.

- Libraries should subscribe to the only database recommended by users.
- Libraries must conduct usability tests for all selected electronic resources before subscription to avoid waste of money and low patronage.

References

[1] Udensi, J. N. & Akor, U. P. (2014). Fundamentals of Library and Information Science. Zaria, ABU Press, 182–186.

[2] Feret, J. (2011). Library as a hub. Changing roles and functions of academic library. A paper presented at the IATUL Conferences. http://docs.lib.purdue.edu/iatul/2011/papers/35.

[3] Tiemo, P. A. & Ateboh, B. A. (2015). Challenges to the development of information and communication technology in University Libraries in Nigeria. In: Nnadi, C., Imhanlahimi, E. O. & Chikwendu, A. V., (ed.), Higher education in Nigeria: perspectives and reforms. Onitsha, Global Academic Group Online Academic Resources, 57–76.

[4] Ekoja, I. I., (2019). Library Awareness Week. Abuja: African Independent Television (AIT) News at 4:00PM.

[5] Manandhar, I., (2019). Evaluating UX: Usability Evaluation. https://blog.prototypr.io/what-and-why-of-usability-evaluation-46bf4b6dee07

[6] Hassan, L. (2014). The website of the university of jordan: usability evaluation. International Arab Journal of e-Technology, 3(4), 258–269.

[7] Quesenbery, W. (2001). What Does Usability Mean: Looking Beyond 'Ease of Use. https://www.wqusability.com/articles/more-than-ease-of-use.html

[8] Nielsen, J. (2012). Usability 101: introduction to usability. Nielsen Group. Retrieved on 10th May, 2019 from https://www.nngroup.com/articles/usability-101-introduction-to-usability/

[9] Joo, S., Lin, S. & Lu, K. (2011). A Usability evaluation model for academic library websites: efficiency, effectiveness and learnability. Journal of Library and Information Studies, 9(2), www.jlis.lis.ntu.edu.tw/article/v9-2-2pdf.

[10] Matera, M., Rizzo, F. & Carughi, G. T. (2006). Web Usability: Principles and Evaluation Methods. In: Mendes, E. & Mosley, N. (eds.), Web Engineering, Berlin, Heidelberg, Springer, 5–18. Retrieved on 14th January 2018 from https://link.springer.com/chapter/10.1007/3-540-28218-1_5#citeas.

[11] Matusiak, K. K. (2012). Perceptions of usability and usefulness of digital libraries. International Journal of Humanities and Arts Computing, 6(1–2), 133–147, Retrieved on 17th December, 2018 from, www.eupjournals.com/ijhac, 10.3366/ijhac.2012.0044.

[12] Lovelock & Wirtz. (2011). Services Marketing-People, Technology, and Strategy, 7th edn. Pearson Prentice Hall, 9–15.

[13] Elkhani, N. & Bakri, A. (2013). Review on Expectancy Disconfirmation Theory (EDT) Model in B2C E-Commerce. Journal of Information Systems Research and Innovation, Retrieved on 15th August, 2018 from http://seminar.utmspace.edu.my/jisri/.

[14] Nigerian Universities Commission. (2019, June 17). List of Approved Universities in Nigeria. Abuja: NUC News Bulletin, 14(24), 1.

[15] Kashyap, J., Khurana, S., Bhandari, P., Gupta, C. & Virmani, D. (2021, March). Webauto_OSSE: University Circulars to Social Media Using JavaScript Open Source Server Environment. International Journal of Emerging Technologies and Innovative Research, 8(3), 1806–1813.

[16] Musyoka, K. (2013). Service quality and library user satisfaction among universities in Kenya. A research project submitted in partial fulfillment of the degree of master of business administration (MBA), university of Nairobi. doi: 103329/ev.v23i/12122.
[17] Lai, P. C. (2017). The Literature review of technology adoption models and theories for the novelty technology. JISTEM, Brazil 14(1), 21–38, www.jistem.fea.usp.br.
[18] Speicher, M. (2015). What is Usability? A Characterization based on ISO 9241-11 and ISO/IEC 25010. Retrieved on 18th September, 2018 from arXivpreprintarXiv, 1–10. In Adepoju, S. A., Oyefolahan, I.O., Abdullahi M. B. & Mohameed A. A. (2018). A Survey of research trends on university websites usability evaluation. A paper presented at the 2nd International Conference on Information and Communication Technology and Its Applications (ICTA) organised by the Federal University of Technology, Minna, Nigeria from September 5 – 6,77.
[19] Agrawal, P., Kaur, H. & Singh, G. (2012). Indexed Tree Sort: An Approach to Sort Huge Data with Improved Time Complexity. International Journal of Computer Applications (0975-8887), 57(18), 26–32, November 2012.
[20] Nigerian Universities Commission (2019, 17th June). List of Approved Universities in Nigeria. Abuja: NUC News Bulletin, 14 (24) 1.
[21] Adepoju, S. A., Oyefolahan, I.O., Abdullahi M. B. & Mohameed A. A. (2018). A Survey of research trends on university websites usability evaluation. A paper presented at the 2nd International Conference on Information and Communication Technology and Its Applications (ICTA) organised by the Federal University of Technology, Minna, Nigeria from September 5–6,77.
[22] Ward, S. M, Freeman, R. S, & Nixon, J. M. (2015). *E-books in academic libraries: stepping up to the Challenge*. West Lafayette: Purdue University Press, Project MUSE. Retrieved from https://muse.jhu.edu/

Anagha Shenoy R, Bhoomika M, Annaiah H

Design of chatbot using natural language processing

Abstract: A chatbot is an artificial intelligence (AI) system that responds to a user's natural language questions with the most suitable answer. The chatbot is an emerging trend that has been set nowadays, to be more precise, during the pandemic. There are many kinds of chatbots based on the principles they work on. Chatbots play a vital role in the interaction with the users who need the information. There are many advantages of implementing a chatbot in any application/website based on the current situation. Numerous chatbots are already deployed and are serving the users, and are striving to fulfill user's needs. Chatbots are mainly made up of many trending technologies, including AI, machine learning, natural language processing, deep learning, and so on, and are made sure that they work up to their best to interact like a normal human being using emerging technologies. The basic architecture of a chatbot is given to acknowledge the working of the chatbot. A case study has been made on the most widely used chatbot – Google Assistant.

Keywords: Chatbot, artificial intelligence, machine learning, natural language processing, pattern matching, natural language toolkit, deep learning, Google Assistant

1 Introduction

A chatbot (chatter-bot) is a technology that allows a computer to communicate with people. The chatbot is known as an artificial conversational entity. Here, conversing through standard text is done using a natural language, like English, and the chatbot responds to the text with the most suitable answer. The first chatbot ever was developed by MIT professor Joseph Weizenbaum at MIT AI Laboratory. Eliza is the first chatbot,made in the year 1966, and its purpose was to give an accurate simulation of a human conversation [1]. It was designed using a simple program to give out predefined responses to the user giving the queries. A test was made to check out whether a program could pass as actual human being capabilities or not. The code was exhaustive to take several possible queries, and the chatbot was capable of passing the test [2]. Today's chatbots have become a lot more advanced than in the past and can answer

Anagha Shenoy R, Dept. Of CSE, Government Engineering College, Hassan, e-mail: anaghashenoyr@gmail.com
Bhoomika M, Dept. Of CSE, Government Engineering College, Hassan, e-mail: bhumikamk12@gmail.com
Annaiah H, Dept. Of CSE, Government Engineering College, Hassan, e-mail: annaiahh@gmail.com

https://doi.org/10.1515/9783110713633-004

very complex queries and have capabilities such as voice interaction where users can ask or interact with the chatbot through their speech. And now, we also have text to speech and speech text integration between user and chatbot. These days' chatbots are not restricted to specific standard languages, as we can see chatbots having multiple language options and few chatbots with local language interaction [1].

1.1 Chatbots and their consequences

A chatbot is an artificial intelligence (AI) system that responds to a user's natural language questions with the most suitable answer. The chatbot is the most promising and advanced interaction between humans and machines as it uses natural language processing (NLP). The purpose of chatbots is to support the user. It is most useful for industries so that with the help of chatbot, they can interact with the customer and know their demands problems and some point even negotiates them with their current product and its feature [1, 3]. Chatbots are made to help the problems that industries are facing today. People have been using messaging apps for interaction with friends and family for a long time, and as they feel comfortable and confident in using these so using the same practices to communicate with a business and technology to effectively communicate through the apps in the way consumers require [1, 2].

1.2 History of chatbots

- ELIZA was the initial chatbot. The term "Chatterbot" was coined in 1994. Joseph Weizenbaum created it in 1966, and it uses pattern matching and substitution methodology to simulate oral communication [3].
- Yank headshrinker Kenneth Colby created PARRY in 1972. The program was for a patient with psychosis. It tries to simulate the sickness [2].
- ALICE was developed in 1995 by Richard Wallace. Unlike ELIZA, NLP was used here, which allowed for more sophisticated conversation. It was open source. Developers could use artificial intelligence markup language to create their chatbots powered by ALICE [1, 2].
- Developer Hrolf Carpenter created Jabberwacky chatbot in 1988. It aimed to simulate natural human oral communication in associate degree entertaining methods [3].
- In the year 2009, a corporation referred to as "WeChat" in China created many advanced chatbots. Since its launch, WeChat has conquered several users. Early in 2016, we tend to see the intro of the primary wave of artificial knowledge technology within the style of chatbots [1].

2 Types of chatbots

1. **Rule-based chatbots**: This can be the best sort of chatbots. Mistreatment choices are unit predefined, and the user moves with these by clicking on buttons. Users should create many picks to induce reply or resolution from the chatbot; these area units are the slowest to guide the client to their goal as these bots have a long user journey. The chatbot asks queries, and folks answer them with a button, and it collects the information and offers a satisfactory reply. But, for many advanced eventualities, these do not seem to be the best option [2].
2. **Intellectually freelance chatbots**: These bots use machine learning (ML) that helps the chatbot learn from the user's inputs and requests. ML can find out by itself from the information, acknowledge patterns, and choose with stripped-down human interference. These chatbots area units are trained to grasp specific keywords and phrases that trigger; they learn and train from expertise [2].
3. **AI hopped-up chatbots**: AI-powered bots mix the simplest from rule-based and intellectually freelance. AI may be a simulation of human intelligence. AI is the space of engineering that focuses on making intelligent machines that work and assume as folks. However, these chatbots perceive free language; however, they even have predefined flows to create optimistic as they solve the user's downside [2].

Based on a specific feature, these three types of chatbots mentioned above are named into different types:

Type 1: **Scripted/quick reply bots** – A scripted chatbot is a pure rule-based chatbot [1].

Type 2: **NLP chatbots** are considered one of the most significant Artificial Intelligence applications (AI). It uses AI technology- NLP (NLP) to get user input [9, 10].

Type 3: **Service/action chatbots** – Service chatbots ask for information from the user to fulfill their request and give a proper reply. This is mainly used in the airline industry. It helps to check flight booking, the cost of reserving the flights, and check the statuses [1].

Type 4: **Social messaging chatbots** – These chatbots make it easy for the customers to directly interact with the bot, just like they do with their friends [1].

Type 5: **Context-enabled chatbots** – These are the most advanced bots. They use ML and AI to remember conversations that happened [1].

Type 6: **Voice-enabled chatbots** – These chatbots accept user inputs through voice [1].

3 Importance of chatbots

Chatbots are the foremost necessary side once it is thought of for business growth because they may gain many customers by having higher user engagement when put next to the archaic ways. It conjointly helps to cut off the delay time which could occur if there have been no chatbots. The chatbots area unit is offered every day, no matter day or night, which makes the work easier with none certain [2]. It is simple to work, too; it simply works like several electronic communication applications. Some chatbots are a unit developed so that they might even speak and reply to the user's queries. The value that to be endowed on a chatbot also will be stripped down, and most come might be expected from it. The fundamental advantage of a chatbot is that it will settle for multiple user queries and answer them simultaneously. Chatbots are significantly quicker and convenient to speak to than any applications used for such purposes [1, 3, 4].

3.1 Significance of chatbots

Chatbots are software robots that reduce human work. Here, the word "work" may refer to the chores, such as answers to the user queries or questions. The implemented chatbot acts as a human interface to the user, and it answers what has been taught with speed and efficiency. It can give answers or replies 24/7 without any change in its efficiency like humans. It can also cut down the workforces required to answer the user's inquiring [1]. It may give out detailed information within no time so that the user's time is saved. Some bots could get updated themselves, and they could work more efficiently without any human intervention. Chatbots could be customized on some themes in which it is designed to respond with emojis instead of text-only. Also, chatbots are used in the segregation of the e-mails using filters for primary, spam, or promotional e-mails [2].

3.2 Advantages of chatbots

- **Available anytime:** The chatbots work 24/7 and are accessible by any part of the world anytime. It works well whether it is day or night [2].
- **Instantaneous replies:** The chatbot comes back with the most relevant answers/solutions for the user-requested queries/questions.
- **Customer engagement and satisfaction:** Most chatbots keep the customers engaged by asking a few questions related to their previous query, and it gives out the most relevant and satisfactory answers to the customer. Hence, it improvises communication and gains more customers [2].

- **Saves time and money:** By implementing the pertinent chatbot, we can save time as it does not require any human supervision for responding correctly, and it could also save the payment of the workforce.
- **Ceaseless endurance:** Chatbots have boundless patience, that is, the user can ask the same question multiple times, and the chatbot will answer it that many times.
- **Manages multiple users:** Chatbots could answer multiple questions from multiple users simultaneously [2].
- **Personal assistant:** They also act as a user's assistant based on their requirements. They can get recommendations for movies, books, music, clothing, and so on, based on the need. This could be done by asking the user about their appropriate choices used during the chatbot training [1].
- **Helps in recruitment:** Many multinational companies use chatbots to select eligible candidates who hold the suitable criteria for that post in the company [2].

These are a few benefits of the chatbot. The main benefit of the chatbots showed up during this pandemic. Due to the outburst of the COVID-19 virus, the whole world was almost shut off. At this time, chatbots were the most important way to maintain the business for some, get out of anxiety for a few, gain knowledge for a handful of people, and many more.
- Chatbots also helped many call centers and many medical organizations to maintain their customer satisfaction.
- Chatbots helped many grocery shops get the order from the user by maintaining physical distance and just delivering their required items.
- Chatbot has also played an essential role in the financial services, the recommendation of movies, books, and songs to those who were quarantined.
- Government and healthcare centers adopted chatbots to reduce their overburdened queries and cases by giving the users practical solutions.
- As the schools and colleges were shut down during this pandemic, chatbots were found very important. They were used in many educational institutes to give their guidelines and notifications.
- Also, the chatbots can assist as a useful resource during the pandemic by sharing some statistics about the number of cases caused by the novel coronavirus not only country-wise but also city/town-wise to create awareness.

4 Real-time applications of chatbots

1. **E-commerce** – Tidio, Sephora Assistant, eBay, Zalando, and so on.
2. **Customer service** – Domino's chatbot, order a Phone Call chatbot, handling returns query bot, Lyft, Slush, Jenny chatbot, PVR cinemas, and so on.

3. **Conversational AI** – Siri, Google Assistant, Alexa, Cortona, AI Responder, Kuki, Visual Chatbot, and so on.
4. **Facebook messenger and telegram messenger** are having options of creating chatbots like Swelly (Facebook messenger), BotMaster (Telegram), and so on.
5. **Gaming bots** – Sire, Amakai, and so on.

These are a few real-time applications. Many more chatbots are being used in day-to-day lives, and many more chatbots are being created based on the requirement.

5 Technologies that empower chatbots

Chatbots are robots that can chat like a human beingwith a software-bot and hence the name chatbot. To make a software robot chat or reply like a human, we might need some technologies like pattern matching, NLP, natural language understanding (NLU), ML, AI, and such technologies software–robot efficient.

Pattern matching, as the name itself says, checks for a specific sequence of characters within the given data. This is of significant use as it takes care of the data entered and helps to get a suitable output based on that pattern [2].

NLP helps achieve the stage where the chatbot would act up like a human during the conversation. NLP is nothing but manipulating the user input, deriving its meaning, and providing the output based on the user input. This is achieved with the help of some steps in NLP, including sentiment analysis, tokenization, name–entity recognition, normalization, dependency parsing, and a few more [5, 6].

NLU is the essential feature of NLP that the chatbot uses to understand the emotions of the user who is interested in getting the information. This feature majorly helps in getting the specific domain of the user's interest. The actions by the chatbot would be taken basically by making use of NLU [2].

ML, as the name itself says about the work, is the learning of the machine. The machine is fed with the data, and the data is used to bring about the required activities using that data. ML or ML helps the machine or the chatbot to acquire some basic knowledge to analyze the user input and reply with the appropriate answer for that query.

AI plays a significant role in giving a human touch to the chatbot in every single conversation. AI becomes a vital component in understanding unique queries, personalizing responses, learning from past conversations (with the help of ML), and using past learning to improvise future conversations.

Furthermore, chatbots might also be developed using artificial neural networks, deep learning, and many more emerging technologies to make them very user-friendly and accurate. **Artificial neural networks** acts as brain cells to the bot to create the network of nodes (neurons), which helps maintain the chatbot's memory. It creates a network-like structure using the nodes to keep up the chatbot's memory and helps to learn using the previous experiences.

Deep learning makes use of ML algorithms to make the software bot act like human-to-human interactions. This helps the users to interact in their native language. Similarly, interoperability, tools for e-commerce, and voice activation are some of the technologies that empower the chatbots [7, 8].

6 Architecture of chatbots

Chatbots are software robots that have to be trained with specific data to reply to the users with the relevant answers for the queries asked by the users. The chatbot architecture could be shown in the Figure 1.

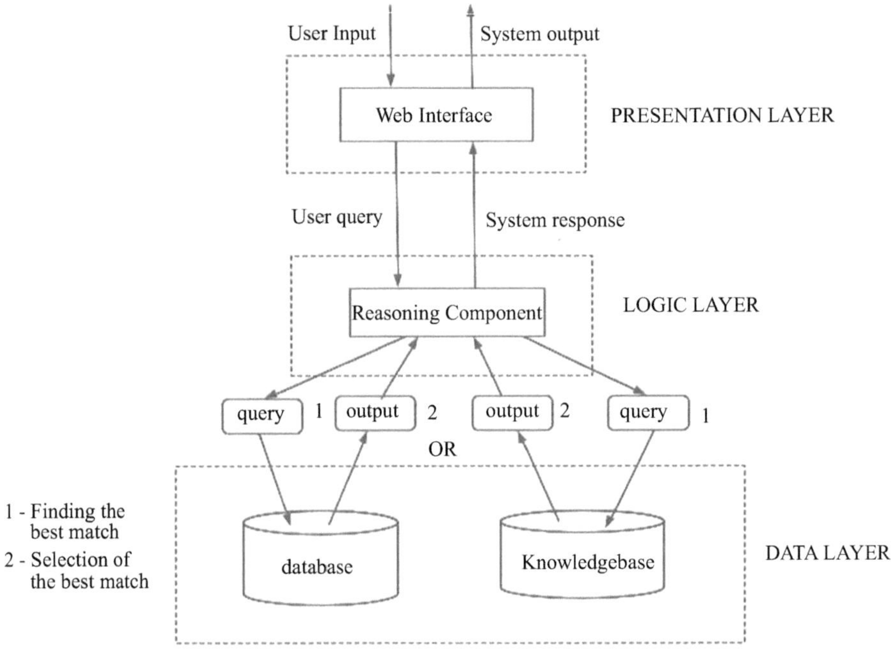

Figure 1: Architecture of chatbot.

As shown in the Figure 1, the chatbot majorly consists of three parts, namely:
1. Web interface.
2. Reasoning component.
3. Knowledgebase and database

The **web interface** supports the front end of the chatbot. Here, the **user input** and **system output** are expected to be given. This is the only interface visible to the user, and the user interactions are to be taken only with this interface.

The **reasoning component** is the intelligent part of the whole chatbot, where the user inputs are analyzed and sent as **user queries** are accepted and then analyzed to give away the most relevant output to the user via the web interface the **system response.**

The **Knowledgebase** is a kind of storage area, where it might be a database or a cloud where all the data could be stored. The reasoning component analyzes the user queries and searches for the best match here, and sends the best result selected.

Basically, the chatbot is categorized into three layers: presentation layer, logic layer, and data layer.

Presentation layer – As the name itself says, it is the layer that is presented to the user, which acts as the graphical user interface, which contains the web interface where the users will interact.

Logic layer – The layer which helps in the working of the chatbot. Here, the reasoning component will be present, which acts as the brain of the chatbot.

Data layer – This layer has all the data that the reasoning component will be used to respond to the user. The Knowledgebase and database are included within this layer.

7 Machine learning

ML is the study of algorithms that, through experience, automatically gets improved. It is also seen as a subset of artificial intelligence. It enables to tackle tasks that people can carry out. From driving cars to translating speech, it is showing the capabilities of artificial intelligence. And for this messy and unpredictable world, it is helping a lot. The process of teaching a computer system how to make accurate predictions when fed data is a very high-level ML definition.

Types of machine learning:

There are two main types of ML are:

i. Supervised learning.
ii. Unsupervised learning

And, we also have other two types named as

i. Reinforcement learning.
ii. Semi-supervised learning.

Supervised learning:
Here, basically, teaching machines are done with the help of an example. Here, during training, the systems are exposed to a large amount of data. These are generally labeled data; for example, images of handwritten figures are fed as data in order to annotate the system to know which number they correspond to. With the help of sufficient examples, the system would learn to recognize the clusters of shapes and pixels. The details associated with each number to distinguish them and some systems needed to be exposed to millions of examples to master a task. Training these systems requires vast amounts of labeled data.

A vast dataset is used to train these systems like with the help of YouTube, its labeled video repository 8 M linking to seven million labeled videos and ImageNet, and Google's Open Images Dataset having about nine million images and also one of the early databases of this kind, having more than 14 million categorized images. The size of training datasets continues to grow with other sites and apps like FB and Instagram.

Example of supervised learning algorithms:
- Linear regression
- Nearest neighbor
- Gaussian Naive Bayes
- Decision trees
- Support vector machine
- Random forest

Unsupervised learning:
Unsupervised learning tasks algorithms identify patterns in data and try to spot similarities by splitting the data into categories. Here, no teacher is required that means no training will be given to the machine. Unsupervised learning algorithms are not designed to single out specific data types, but they simply look for data that similarities can group. Here, the machine is restricted to find the hidden structure in unlabeled data.

Unsupervised learning classified into two categories of algorithms:
- **Clustering**: To discover the inherent groupings in the data, such as grouping customers by purchasing behavior, is a clustering problem.
- **Association**: An association rule learning problem is where people that buy X also tend to buy Y here; we try to discover rules that describe large portions of your data.

Types of unsupervised learning:

Clustering
1. Agglomerative
2. Overlapping

3. Exclusive (partitioning)
4. Probabilistic

Clustering types:
1. Hierarchical clustering
2. Principal component analysis
3. Singular value decomposition
4. Independent component analysis
5. K-means clustering

Semi-supervised learning
This approach is a mixture of supervised and unsupervised learning. It relies upon using a small amount of labeled data and many unlabeled data to train systems. The enormous sets of labeled data for training machine-learning systems importance are diminishing over time due to the rise of semi-supervised learning.

Reinforcement learning
To understand reinforcement learning, let us think about how someone might learn to play an old-school computer game for the first time when they are not familiar with the rules or how to control the game. While they are playing, eventually, they learn. Their performance will get better and better. The process of many cycles of playing the system builds a model of which actions will maximize the score.

Implementation
ML chatbots are already interacting with consumers. The main types of dialogue systems are goal-oriented ones like Siri and general conversation ones like Zo, which is now retired. Business chatbots can supply our customers with the same level of attention they would get in a physical store and also guide them through the entire shopping experience with the help of a live chat interface.

Let us take an example of Amazon-Alexa, which is goal-oriented. *Alexa is one of the most famous examples of conversational AI.* The goal-oriented dialogue system exists to answer specific questions and to perform functions like switching video or changing different music. The second is based on the ELIZA bot of 1960s, which is still available to chat with online today. ELIZA would not be able to switch video or change different music. Here, users converse with it to test the limits of the bot. Let us imagine a conversation with a bot. The conversation, in this case, is centered on geographic allocation.

User: hi

Bot: Hello! Welcome, how may I help you?

Now, this is just a simple example of how ML could work. Here, some bots are designed for specific purposes where you can ask only those questions. For example, you are designing a bot for your college, and you ask something regarding movies, it will not be able to answer because that bot is not trained for that. During the development, the phrases are designed to help the bot understand the context and infer responses from user inputs.

Every query and every statement we enter into the chat box is helping the bot to become more intelligent and more efficient the more it is used and indirectly getting trained. The more we use, the more familiar we become and also get to know the problems too. And depending on the response, new features can also be added.

Chatbots with machine learning: building neural conversational agents
Interacting with the machine with the help of natural language is one of the requirements. This field is called chatbots as we expect the machine to provide us an informative or most accurate answer, which is not more different from the human. That means the answer or the response must be accurately equal to the typical human response for that particular question.

There are two significant types of dialogue systems: ***goal-oriented*** *and* ***general conversation****. Goal-oriented is Siri, Alexa, Cortana, and so on, and general conversation is Microsoft Tay bot.*

This helps people to solve everyday problems using natural language, while the latter attempts to talk with people on a wide range of topics.

8 Natural language processing

NLP is a subdivision or a part of AI, which fundamentally deals with the interaction between the human and the computer by using natural language. The superlative objective of NLP is to read, decrypt or extract, understand, and be coherent of the language that humans will use to interact with the machine in such a manner that it seems that the human interacts with another human instead of the human interacting with the computer. Primarily, NLP is being used for interactions such as a human interacting with the machine; machine that would be able to catch and reproduce audio; text-to-voice and voice-to-text conversion; processing of text that will be input into the machine; majorly, a machine that interacts with the human with audio along with the text and many more. Nowadays, NLP is also used so that a dummy human video is placed that interacts with the users of the chatbot and acts as human-to-human interaction instead of human-to-machine interaction. Translation of languages like Google Translate, word processor applications such as Grammarly or Microsoft Word for checking grammatical accuracy of the texts that are typed in those applications are some significant applications of NLP. They can be found in call centers to automate the response to user's queries such as interactive voice response and can extensively be used in applications that act as personal assistant such as Google Assistant, which responds to the user when the user utters OK Google; Siri, a personal assistant in iOS devices; Cortana; Alexa; and more.

NLP uses deep learning that the computer uses to acquire the actual meaning of the user's inputs and give responses and then respond to that query accurately. Using NLP, the chatbot must be trained on multiple interactions, and also, the application will be able to learn upon its previous interactions.

The main techniques of NLP are **syntax analysis** and **semantic analysis**. Let us consider the syntax analysis; the sub-techniques are –

- Lemmatization – a grouping of similar kinds of words so that they are classified and processed together.
- Morphological segmentation where the words are divided into morphemes (individual unit).
- Word segmentation takes place to divide a continuous sentence into parts.
- Parsing, checking for grammatical errors for the selected sentence.
- Identifying parts of speech in the sentence and check for errors, if any.
- Breaking of the sentence into parts based on keywords.
- Stemming – removing the inflected words from the sentence, often from the end part of the sentence.

Semantic analysis is the process that checks for the correctness of meaning in a sentence; the sub-techniques are –

- Named entity recognition – as the name itself says, it includes detecting the text parts which can be identified and categorized into groups of similar types.
- It involves the disambiguation that involves giving sense or meaning to a word or phrase.
- It also fetches from the database after semantic checking and converts them into human language.

Implementation using natural language processing:

The commonly used technologies for developing chatbots are:

1. Python – a high-level, dynamically typed programming language which helps in building the chatbot architecture.
2. Pandas – a software library used in Python language to manipulate data and analyze the input data.
3. TensorFlow – a software library that is used in the implementation of neural networks and ML.
4. SpaCy – an open-source library software that is used in advanced NLP.
5. Application program interfaces are required to connect the chatbot application to the websites or any messaging platforms.

Natural language tool-kit

Natural language tool-kit (NLTK) is a platform used for building programs in Python, which would work with data in human language. This is a fantastic tool used to teach a machine a language used by humans for interaction in day-to-day life using Python. It is a software library that is used in Python, which is used in making the computer understand natural languages. NLTK was developed with four goals, namely – **simplicity**, **consistency**, **extensibility**, and **modularity**.

To download and install NLTK:

- To install the NLTK, run the given command:
  pip install nltk
- To check whether the toolkit is installed correctly or not, follow:
  Run python3, then type, import nltk

To install NLTK packages:

import nltk and Run nltk.download()

The above shown would open the NLTK package downloader from where we can select the corpora or models required and download them or download all of them.

To pre-process the text with the help of NLTK:

The data that would be given into the application will be in the form of string or text. The input text will not be in the proper format, which has to be fed into the application. In order to make to application an acceptable format, we might have to perform a few pre-processing of the text. *Basic steps of text-processing include:*

- As the given input will be having the text having a mixture of upper case and lower case letters, it would be tough to make the machine understand as it treats the exact words in different cases as different ones. So, converting the whole text either into uppercase or into lowercase so that the algorithm treats all of them identical.
- A sentence tokenizer and a word tokenizer are used to mark few critical keywords in the text. Tokenization is the process that is used to convert the text into a list of tokens. Here, tokens are the keywords.
- Removal of noise has to be done. Here, "noise" means the letters or numbers which are not in a standard format.
- Erasing of the stop words. Stop words are the common words that are present and less useful in considering the user query. These words have to be excluded.
- Stemming – removing the inflected words from the sentence, often from the end part of the sentence.
- Lemmatization – a grouping of similar kinds of words so that they are classified and processed together.

Bag of words:

After the pre-processing of the text now, we have to put those strings into a meaningful array. A Bag of words (BoW) is a method of extracting from the strings for using them in modeling with the ML algorithms [12]. Fundamentally, a BoW represents strings that tell us the number of occurrences in the text-processed document. BoW involves two things:

i. A vocabulary of known words.
ii. A measure of the presence of known words.

For example, if we consider a dictionary that contains the words {Studying, is, the, not, good}, and we want that to vectorize it to "Studying is good," we would have the vector as: (1, 1, 0, 0, 1).

Term frequency–inverse document frequency:

As there is a problem with the BoW approach, we are finding the term frequency–inverse document frequency (TF-IDF). The problem with the Bag of words approach is that the words which appear more times in the document start to dominate the document, and this might not have the actual informational content. Also, this approach gives out more preference to the longer document instead of, the shorter document.

To avoid such issues, we use the TD-IDF, where the frequency of the words will be rescaled by how many times they have repeated so that the words like "the" are going to be penalized.

Term frequency is the score of the count of the frequency of the word in the current document.

$$\text{TF} = \frac{\textit{Number of times 't' term appears in the document}}{\textit{Number of terms in the document}}$$

IDF is the score of noting the rareness of a word in the whole document.

$$\text{ITF} = 1 + \log\left(\frac{N}{n}\right)$$

where "N" is the number of documents and "n" is the number of documents

TF-IDF weight is used in the retrieval of information and text mining. This weight is used to evaluate how important a word is in a document within a collection [12].

*For example, consider a document that contains 100 words, wherein the word "college" appears ten times. The TF for college is (10/100) = 0.1. Now, let us assume 10 million documents, and the word "college" appears 1000 times. The IDF for college is (10,000,000 / 1000) = 4. Thus, the TF-IDF weight is the product of these quantities: 0.1 * 4 = 0.4.*

Cosine similarity

TF-IDF is a transformation that will be applied to the text to get two vectors which are real-values. Cosine similarity is a measure of the similarities between two non-zero vectors. Cosine similarity of the pair of vectors can be obtained by taking the dot product of the vectors and dividing them by the product of the norms of the vectors. This gives us the cosine of the angle between the vectors. The formula is given as:

$$\text{Cosine similarity (d1, d2)} = \frac{\textit{Dot product}\ (d1, d2)}{||d1|| * ||d2||}$$

Where, d1 and d2 are non-zero vectors [12].

1. The libraries that have to be implemented

```
import nltk
import NumPy as np
import random
import string
```

2. Corpus – the document that has the actual content

This is the document that contains the raw data. Create a text file named "chatbot.txt" within which all the data must be kept [12].

3. Reading the data

The data in the corpus will be read and will be converted into a list of words and sentences which can be used for pre-processing [12].

```
f= open('chatbot.txt', 'r',errors = ignore')

raw = f.read()
raw = raw.lower

nltk.download('punkt')
nltk.download('wordnet')

sent_tokens = nltk.sent_tokenize(raw)
word_tokens = nltk.word_tokenize(raw)
```

4. Pre-processing of the raw text

Let us define a function named LemTokens, which takes the input which is token and gives out the normalized tokens [12].

```
lemmer = nltk.stem.WordNetLemmatizer()

def LemTokens(tokens):
    return [lemmer.lemmatize(token) for token in tokens]
```

```
remove_punct_dict = dict((ord(punct), None) for punct in string.punc-
tuation)
def LemNormalize(text):
    return
LemTokens(nltk.word_tokenize(text.lower().translate(remove_punct_
dict)))
```

5. Matching of the keyword

Here, we will define a function that greets the user if the user's input is a greeting [12].

```
GREETING_INPUTS = ("hello", "hi", "greetings", "sup", "what's up","hey",)
GREETING_RESPONSES = ["hi", "hey", "*nods*", "hi there", "hello", "I am
glad! You are talking to me"]

def greeting(sentence):
  for word in sentence.split():
    if word.lower() in GREETING_INPUTS:
       return random.choice(GREETING_RESPONSES)
```

6. Generation of response

To generate the response to the user query, these steps could be done:

– Importing TfidfVectorizer from sci-kit learn library, which is used to *convert a collection of raw documents into a matrix of TF-IDF features*: [12]

```
from sklearn.feature_extraction.text import TfidfVectorizer
```

– Importing cosine_similarity from sci-kit learn library, which is used to *find the similarities between the words that user have entered and the words that are present in the corpus*: [12]

```
from sklearn.metrics.pairwise import cosine_similarities
```

– When the user enters the query which is not related to any keywords in the consent, it must return a message which says, "I am sorry! I don't understand you" [12]

```
def response(user_response):
    robo_response=''TfidfVec =
TfidfVectorizer(tokenizer=LemNormalize, stop_words='english')

    tfidf = TfidfVec.fit_transform(sent_tokens)
    vals = cosine_similarity(tfidf[-1], tfidf)
    idx=vals.argsort()[0][-2]
    flat = vals.flatten()
```

```
    flat.sort()
    req_tfidf = flat[-2]

if(req_tfidf==0):
        robo_response=robo_response+"I am sorry! I don't understand you"
        return robo_response
    else:
        robo_response = robo_response+sent_tokens[idx]
        return robo_response
```

- The response that the application must give while beginning and ending the conversation must be given as: [12]

```
flag=True
print("ROBOT: My name is ROBOT. If you want to exit, type Bye!")

while(flag==True):
    user_response = input()
    user_response=user_response.lower()
    if(user_response!='bye'):
        if(user_response=='thanks' or user_response=='thank you'):
            flag=False
            print("ROBOT: You are welcome..")
        else:
            if(greeting(user_response)!=None):
                print("ROBOT: "+greeting(user_response))
            else:
                sent_tokens.append(user_response)

word_tokens=word_tokens+nltk.word_tokenize(user_response)
                final_words=list(set(word_tokens))
                print("ROBOT: ",end="")
                print(response(user_response))
                sent_tokens.remove(user_response)
   else:
       flag=False
       print("ROBOT: Bye! take care..")
```

This is a basic chatbot that will work fine if the user asks any queries with the keyword present in the corpus text file. The chatbot here is named as ROBOT [12].

Things a developer must keep in mind while developing a chatbot:
- The chatbot must be affordable.
- Security of chatbot.

- User-friendly design so that people like your chatbot.
- Choosing between voice-activated and text-based chatbots.
- Making sure chatbot provides real value to use.
- Alternate solution to problems that may arise in future in order that coping up becomes easy when things get it wrong.
- Knowing when a human needs to take over the machine.
- Understanding how successful the chatbot should be.

Case study

Google Assistant

The virtual assistant that every android phone powered by Google has is the Google Assistant, which would do most of the jobs, including setting the alarm, sending messages, calling a person, and many more just through a single command to the bot.

Google Assistant is a chatbot that merely uses AI developed by Google developers. This is primarily made available in all the android devices powered by Google and few smart home devices. The Google Assistant purely acts as a virtual assistant obeying the users' commands and acts as an interactive machine that replies to the user queries like a human [10].

Initially, Google Assistant was introduced in Google's messaging application called "Allo" in May 2016. Later on, it was implemented in Google Home, a voice-activated speaker, and then exclusively on Pixel and Pixel XL. The deployment of this virtual assistant in all android devices started in February 2017.

Google Assistant is developed so that it can be just accessed with a normal human voice tone (yes, it is developed to be that sensitive) and through the keyboard. It can do multiple tasks if permission is granted on android devices. It majorly runs as a mini-version of the search engine. Apart from this, this assistant would also make changes in the phone's hardware, set the alarm, make a phone call, send a message, set a reminder, make a shopping list, and also chat with the user in his native language. It would also get the user a joke, or give away any facts from the internet or even play games upon the user's request [10].

Nowadays, apart from mobile phones, it is also being used in android televisions, tablets, smart home devices, and so on. The first Google Assistant had made its debut in Android Marshmallow or Nougat, which could speak in English. Later on, it was integrated into Android Wear 2.0. The first-ever laptop that had Google Assistant was Google Pixelbook in December 2017.

Google Assistant uses the NLP algorithm, developed in Google itself, which majorly tends to bring about the one-to-one human-like conversation with the user and give the same reply to the user query. Some more updates were done, and the chatbot was trained to understand and speak up of over 30 languages in more than 90 countries [10].

The essential feature of the Google Assistant includes

- Letting third-party device makers incorporate the actions on Google commands for their respective products.
- Incorporating text-based interactions.
- Incorporating more languages.
- Allowing users to set a geographic location precisely for the device to enable improved location-specific queries.

9 Conclusion

A chatbot is a software application used to conduct an online chat conversation instead of providing direct contact with a live human agent. Its creation and implementation is a developing area related to AI and ML, so the provided solutions, while possessing advantages, will be having a few critical limitations in terms of functionalities and use cases. However, this is changing over time. There are few disadvantages like the limited data to reply; there might be few mistakes that could be made while training the bot, which will highlight in the responses, it will require a strong internetwork connection which might also be considered as a disadvantage. Also, some obsolete bots, working on the old generation technology, needs to be replaced. The chatbots might also be created to acquire the user's data, so the user must look after the type of chatbot, whether it is from a reliable source while interacting with. Hence, we may conclude that the chatbot has a slight disadvantage and a high number of advantages, the creator and the user must make practical use of chatbots.

References

[1] Anagha Shenoy, R., Bhoomika, M., Annaiah, H. A Survey on Chatbot and its relevance in present-day, Recent Developments in Engineering and Technology-2020, Page: 275–288.

[2] Anagha Shenoy, R., Bhoomika, M., Annaiah, H. Chatbots and their relevance in present situation, Review of Language Literature, Science, Commerce and Humanities, Page: 278–283.

[3] Dale, R. (2016, CrossRef Google Scholar and). The return of the chatbots. Natural Language Engineering, 22(5), 811–817., kevincurran.org.

[4] Eleni, A., Lefteris, M. An Overview of Chatbot Technology.

[5] Gupta, C., Jain, A., Joshi, N. (2018, Jan 1). Fuzzy logic in natural language processing–a closer view. Procedia Computer Science, 132, 1375–1384.

[6] Gupta, C., Jain, A., Joshi, N., Novel, A. Approach to feature hierarchy in Aspect Based Sentiment Analysis using OWA operator. In Proceedings of 2nd International Conference on Communication, Computing and Networking 2019 (pp. 661–667). Springer, Singapore.

[7] Gupta, C., Chawla, G., Rawlley, K., Bisht, K., Sharma, M. Senti_ALSTM: Sentiment analysis of movie reviews using attention-based-LSTM. In Proceedings of 3rd International Conference on Computing Informatics and Networks: ICCIN 2020 (p. 211). Springer Nature.
[8] Gupta, C., Jain, A., Joshi, N. (2019, Jan 1). DE-ForABSA: a novel approach to forecast automobiles sales using aspect based sentiment analysis and differential evolution. International Journal of Information Retrieval Research (IJIRR), 9(1), 33–49.
[9] Verma, P., Agrawal, P., "Study and Detection of Fake News: P2C2 Based MachineLearningApproach", 4th International Conference on Data Management, Analytics & Innovation (ICDMAI), 261–278, New Delhi, Jan 2020, Springer. https://doi.org/10.1007/978-981-15-5619-7_18
[10] Agrawal, P., Kaur, H., Kaur, G. (2012, November). Multi lingual speaker identification on foreignlanguages using artificial neural network. International Journal of Computer Applications (0975–8887) Volume, 57(13), 36–42.
[11] Jain, L., Agrawal, P. (2017, Feb). English to Sanskrit transliteration: An effective approach to design natural language translation tool. International Journal of Advanced Research in Computer Science (IJARCS), 8(1), 1–10.
[12] Agrawal, P., Madaan, V., Sethi, N., Kumar, V., Singh, S. K. (2016). A novel approach to paraphrase english sentences using natural language processing. International Journal of Control Theory and Applications, 9(11), 5119–5128.
[13] An Overview of Chatbot Technology (Accessed on November, 2020 from https://www.researchgate.net/publication/341730184_An_Overview_of_Chatbot_Technology).
[14] Chatbot Examples (Accessed on August 2020, https://botsociety.io/blog/2018/03/chatbot-examples/).
[15] Chatbot (Accessed on August 2020, https://en.wikipedia.org/wiki/Chatbot).
[16] Google Assistant (Accessed on November 2020, https://assistant.google.com/).
[17] What are bots (Accessed on July 2020, https://botscrew.com/blog/what-are-bots/).
[18] Natural Language Toolkit (Accessed on November 2020, https://medium.com/@ritidass29/create-your-chatbot-using-python-nltk-88809fa621d1).

Neeraj Bhanot, Parth Padalkar

Algorithm development based on an integrated approach for identifying cause and effect relationships between different factors

Abstract: Cause–effect analysis is crucial for studying any complex system with various factors affecting the output. The software aims to provide a more straightforward approach for cause–effect analysis on many factors in any chosen system. This has been made possible by integrating the algorithm of special decision-making techniques, that is, "decision making trial and evaluation laboratory," "maximum mean de-entropy," and "interpretive structural modeling" into one program. The software takes the number of factors and their relation as input and gives the final reachability matrix and a level-wise directed graph of all factors portraying the most dependent factor at the top and most driving factor at the bottom. This software application is immense and can be found in industries ranging from production to IT domain. This software will be particularly very useful for management scholars in their research.

Keywords: Python framework, cause–effect analysis, DEMATEL, MMDE, ISM

5.1 Introduction

There are numerous examples in the available literature of decision-making techniques being employed to evaluate the results of various problems in a variety of contexts. Decision-making techniques have been in use for a long time, and they are generally used for establishing the interrelations between different factors in a system. Approaches such as "decision making trial and evaluation laboratory" (DEMATEL), "maximum mean de-entropy" (MMDE), and "interpretive structural modelling" (ISM) are beneficial in identifying the hierarchical significance of the factors in terms of their power to affect each other as well as the system which they comprise.

Kumar and Dikshit [1] used these techniques to find barriers to Waste of Electrical and Electronic Equipment (WEEE) management; Zhao, Chen, and Li [2] analyzed and evaluated the relationship among factors that affect "Renewable Energy Power Generation Projects" (REPGs) development in China; Kapse et al. [3] showcased the major driving factors responsible for the entrepreneurial inclination of people toward

Neeraj Bhanot, Scientist, CSIR – National Physical Laboratory, New Delhi, India – 110012, email: neeraj.bhanot@nplindia.org
Parth Padalkar, PhD Student, The University of Texas at Dallas, Texas, USA – 75080, email: parth.padalkar@utdallas.edu

https://doi.org/10.1515/9783110713633-005

the textile industry. Jain and Raj [4] portrayed a structural model of the various performance factors associated with flexible manufacturing system (FMS) with the help of these techniques; Govindan et al. [5] used these techniques to identify the factors that were most critical for identifying the best "third-party reverse logistics provider" (3PRLP). Such examples show that the contexts in which these techniques can be applied are infinite.

Graph-theory-based DEMATEL technique divides the identified factors into the cause and effect groups to understand their interdependence and tackle the problems visually. It can further help in developing a directed graph to highlight interrelationships between different factors. Using the MMDE method, a threshold value is set for obtaining the impact relations map. The traditional method is conducting discussions with experts and agreeing upon a typical threshold value. The work involved in obtaining a threshold value using the traditional method becomes more complex as the number of factors increases. This algorithm uses the entropy approach to derive a set of dispatch nodes with a strong capacity to influence others and receive nodes easily influenced by others. Based on the obtained sets, a threshold value can further be obtained to map the impact relations. ISM is a technique used to create graphical representations of complex systems.

The calculations involved in implementing these techniques are immensely complicated, lengthy, and error-prone. Attempting to calculate manually could lead to errors that may have cascading implications on the subsequent steps of the calculations. It would take a lot of time and effort of the person implementing these techniques, and the problems generally demand exact calculations, thus adding to the hardships of the person.

Thus, an attempt has been made to minimize the efforts and time required to implement these techniques for identifying interrelationships between factors of any particular area in concern by creating software using the python programming language. It is an interpreter, high-level, general-purpose programming language used by developers worldwide to create interactive software. Guido van Rossum had developed and released it in 1991. This language has a philosophy that emphasizes code readability. Thus, the software will automate the calculations for the user and give exact results in very little time. It is designed in such a manner as to integrate the three abovementioned techniques (DEMATEL, MMDE, and ISM) into a straightforward program that would take the users' entered data as inputs and give the individual results after implementing the techniques on the data as outputs. In this software, ISM has been used to obtain a directed graph based on the interdependency of various factors.

Through this software, an effort has been made to use a few decision-making techniques in conjunction with one another to perform cause–effect analysis of a complex system influenced by various factors by obtaining a directed graph of the interrelationships of those factors.

5.2 Applications of cause–effect analysis oriented studies in different contexts

This section highlights the applications of DEMATEL, ISM, and MMDE techniques and their significance while showcasing their use in obtaining systematic results from problems having a vast number of dependent and independent factors.

Kamble et al. [6] used ISM to establish interrelationships between barriers while adopting industry 4.0 practices in the manufacturing sector within the Indian context. These results were further used as inputs to fuzzy MICMAC analysis to determine the driving and dependence power of those 12 barriers. The results of the study compartmentalized the considered barriers into four categories vis. a vis. linkage, autonomous, dependent, and independent ones to develop the theory of Industry 4.0 adoption in manufacturing organizations. Govindan et al. [5] also used the ISM approach for the relationship analysis between specific factors critical for identifying the best 3PRLP from "n" 3PRLPs. All the attributes were grouped into seven main attributes, and a digraph of their interrelation was obtained. Based on the driving power and dependence, the attributes were further classified using MICMAC analysis into four categories: "Autonomous, Dependent, Linkage, and Independent." The selection of the best 3PRLP was necessary because it has a significant impact on the company's performance. Jain and Raj [4] used ISM to create a structural model of the various levels of the performance factors associated with FMS based on their interrelationship. The dependence and driving power of the factors were also evaluated by separating the factors into clusters, namely, "Autonomous, Dependent, Linkage, and Independent." The study identified three factors: "quality, productivity, and flexibility," which significantly affect the performance of FMS. Mangla et al. [7] considered the 10 most significant sustainability enablers to implement sustainable initiatives in the agri-food supply chain after rigorous research. They used the ISM and fuzzy DEMATEL technique for their analysis. The ISM technique provided the interrelationship between the various factors as well as their driving and dependence potential. In contrast, the DEMATEL technique helped them classify the enablers as influential and influenced while also categorizing them into cause and effect groups. "Pressure by various governments, regulating agencies and non-government bodies" and "understanding customer and other stakeholder requirements" were found out to be the bottom level enablers with the highest independent powers. DEMATEL method helped categorize six of the enablers as the cause group and the other four enablers as the effect group. Zhang et al. [8] used the DEMATEL technique to determine the interrelationship between various risk factors associated with Sponge City public-private capitalhip (PPP) projects in China to analyze and identify the critical risks. Eighteen risk factors were considered and their cause–effect relationship was evaluated to find out the critical risks while generating a scatter plot of the cause degree. The study showed that "inadequate supervision system, government intervention, immature law and regulations, project fragmentation and unclear boundary of catchment areas" were some of the most critical risk factors, which needed strategic attention. Zhou et al. [9] used the ISM technique to identify the drivers toward the Chinese ELV recycling business from the perspectives of the government, recycling organizations, and the consumers. The aim was to provide some strategic insights to improve the sustainability of the automobile supply chain. Different ISM-based graphs showing the interdependence of various drivers were created for government, industrial organizations, and individual consumers' perspectives. The drivers were further categorized into clusters, namely: "autonomous, dependent, independent, and linkage" based on their driving and dependence power using MICMAC analysis for all three perspectives. The results showed that regulations on auto-factory, disassembly technique, and value mining of recycling business, if improved, will improve ELV recycling in the Chinese market. Raj et al. [10] identified the enablers for transition to FMS in the Indian market context and analyzed them through the ISM technique to find out the dependent and the driving enablers and evaluate their interdependence. Eighteen enablers were identified and were structurally arranged into different levels according to their driving and dependence power. MICMAC analysis was also used to categorize the enablers into four categories: Autonomous, Dependent, Independent, and Linkage. The results of this study showcased the most important attribute to shift from traditional system to FMS as the vision and commitment of the top management while the enablers which have higher driving power need to be given higher priority which is "top management commitment, clear vision, and Effective long term planning." Kapse et al. [3] identified 22 factors that contribute towards the entrepreneurial inclination of people to the textile industry. ISM technique was used to create a structural model of the interrelationships of the factors hierarchically. Further, MICMAC analysis was used for finding out the driving force factors and the dependence power. The factors were divided into "autonomous, dependent, independent, and linkage" based on the driving and dependence power. The study showed high driving power for entrepreneurial policy-related factors. Also, the institute's head was found to be an emerging factor responsible for the students' propensity towards entrepreneurship. Panahifar et al. [11] identified 45 barriers and their interrelationships to "collaborative planning, forecasting, and replenishment" in the context of high-tech industries using the ISM technique. Furthermore, MICMAC analysis was utilized to evaluate the driving and dependence power of the factors. The study identified "lack of visible and effective leadership" as the most dominant and "lack of technical expertise" as the second most dominant barrier. Kumar and Dixit [1] identified seven primary and 44 sub-barriers to "WEEE" management in the context of India. They used the DEMATEL method to evaluate the interdependence of the barriers. Various causal diagraphs were generated and the cause–effect relationships were evaluated based on the driving and dependence powers of the factors. The results thus obtained from the study showcased that policy and regulatory barriers and socio-economic barriers have a maximum driving effect on the overall system.

Lin et al. [12] proposed an improved ISM methodology to calculate the correlation coefficient between the influencing factors using grey relational analysis. This was done to analyze the factors which influence food safety hierarchically. A multi-hierarchy structure was stratified and established for the factors influencing food safety using ISM. Zhao et al. [2], through a literature survey, identified 16 representative factors that affect the development of "REPGPs" in China. ISM technique is utilized to obtain a hierarchy of interrelationships of the representative factors. The factors are further classified into four clusters using MICMAC into "autonomous, dependent, independent, and linkage." The study showcased that "economy and urbanization development, incentive policy system, and government policy implementation" are the most significant factors. This would aid the policymakers in understanding the impact of forces on the development of REPGPs. After rigorous literature review, it was noted that these techniques had been used in various contexts like determining the enablers and barriers for a particular instance concerning the Indian manufacturing context, agri-food supply chain context, businesses, increasing entrepreneurship inclination, markets, checking food safety, and so on. These techniques are also used widely for determining the structural hierarchy and interrelationships between various factors, as seen in other published studies [1, 4, 6].

Although these techniques are widely used in various studies, there seems to be a lack of an integrated tool or software that would automatically implement these techniques on the input data given by the user and return the individual results as outputs. The lack of such a tool is a significant hindrance in carrying out the calculations involved in the implementation of these techniques. The requirement of laborious calculations about these techniques gives room for error and takes a lot of time and effort for the user if it is carried out manually. Thus, recognizing the need for a tool to automate these techniques and reduce the margin for error to minimal, we developed software that integrates DEMATEL, ISM, and MMDE techniques into a simple, user-friendly program that takes in the relevant data as inputs and gives the outputs as required with zero error.

5.3 Problem statement

The implementation of decision-making techniques and cause–effect analysis of a complex system comprising various interdependent factors requires many manual calculations. This is the main problem that has been addressed and whose solution has been proposed by automating the required techniques for the cause–effect analysis in Python code. The software hence developed, incorporates the algorithms used for DEMATEL, MMDE, and ISM into one program that can be used to perform the analysis on any system. The data utilized in this chapter has been considered from one of our recently published research papers focusing on "An integrated approach for analyzing the enablers and barriers of sustainable manufacturing" [13]. It uses the same techniques to find the degree of impact of the enablers and barriers to sustainable manufacturing on sustainable manufacturing itself. The problem statement and efforts have been made to evaluate the results using the software that automates the same techniques. The results then obtained by using the software are verified with the results cited in the paper. The list of enablers considered for study has been presented as follows:

1. **Pressure from the market (E1):** The leading indicators of pressure from the market are the trade and commercial practices, competitors, and customer satisfaction. This pressure is on the manufacturing organizations to stay competitive in business.
2. **Government promotions and regulations (E2):** Government promotions and regulations play an essential role in determining the practices and techniques that the manufacturing organizations adopt. The government can impose laws on the use of specific techniques to promote sustainable manufacturing.
3. **Economic benefits (E3):** The economic benefits or profit gained by the manufacturing organizations by implementing techniques that promote sustainable manufacturing is a crucial factor.
4. **Investment in innovation and technology (E4):** The manufacturing organization's inclination toward innovation, especially in sustainable manufacturing, should be steep as it leads to more profit and better living conditions.
5. **Lowering manufacturing cost (E5):** As the manufacturing cost is reduced, there is more investment opportunity in techniques that favor sustainable manufacturing.
6. **Improving quality (E6):** Improving the standard of quality serves as a powerful enabler to sustainable manufacturing as the improved quality leads to more demand and profits and increases the competition.
7. **Education and training system (E7):** With better education and training for the employees, safety in the work environment is improved.
8. **Attractive foreign direct investment (E8):** Increasing foreign direct investments leads to better technology, improvement of the marketing network, and healthy competition. This increases the opportunities to adapt sustainable manufacturing techniques.
9. **Infrastructure facilities in the transportation sector (E9):** This refers to developing the infrastructure for viable air, road, and rail connectivity, which leads to greater competition.
10. **Development in E-economy (E10)**: India has an excellent reputation for delivering information and communication technology (ICT)-related services. ICT can act as a powerful tool to improve technology and subsequently improve the competitiveness of the manufacturing sector.

5.4 Methodology

Three decision-making techniques have been integrated and used in conjunction to get the required final reachability matrix and directed graph showing interrelationship between various factors entered by the user.

The tool has been developed such that the input it takes is an $N \times N$ matrix where N is the number of factors that need to be examined. The row values of each factor determine the impact it has on influencing the corresponding column factor.

DEMATEL technique is then implemented as follows:

Step 1: Questionnaire formulation and response collection.

A questionnaire is prepared, and the responses of a group of academicians and industrialists are noted in the form of an N x N non-negative matrix wherein each individual gives a value in the interval [0, 4] to rate the factor impact degree in a row "i" has on the factor in a column "j." The values 0 to 4 represent the following meaning:
0 – "No Influence."
1 – "Low Influence."
2 – "Medium Influence."
3 – "High Influence."
4 – "Very High Influence."

Step 2: Calculation of average matrix "**A.**"

If the pair-wise comparison between two factors in concern is represented by x_{ij}^{k} then the $N \times N$ matrix is represented by $X^k = x_{ij}^k$ with k = 1, 2, 3, . . ., h; where h is no. of participants with diagonal elements of each matrix set to 0. The $N \times N$ average matrix is calculated by averaging all the matrices as follows:

$$a_{ij} = \frac{1}{h}\sum x_{ij}^{k} \tag{5.1}$$

Step 3: Calculation of direct influence matrix.

The average matrix is further utilized to calculate normalized direct-relation matrix "**D**" as follows:

$$s = max\left(max_{1 \le i \le n} \sum_{j=1}^{n} a_{ij}\,,\; max_{1 \le j \le n} \sum_{i=1}^{n} a_{ij} \right) \tag{5.2}$$

Hence,

$$D = \frac{A}{s} \tag{5.3}$$

As the summation of each row "j" of matrix "**A**" represents the direct impact of one factor over others, $max_{1\leq j\leq n} \sum_{j=1}^{n} a_{ij}$ gives the factor executing the highest direct impact over others. Similarly, the summation of each column "*i*" of matrix "**A**" gives the direct impact received by factors; therefore, $max_{1\leq j\leq n} \sum_{i=1}^{n} a_{ij}$ gives the factor which is majorly influenced by other factors.

Step 4: Calculation of total relation matrix.

The matrix "**D**" is used to calculate the total relation matrix "**T**" as follows:

$$T = t_{ij} = (I - D)^{-1} \tag{5.4}$$

where I is an $n \times n$ identity matrix.

Step 5: Calculation of rows and columns summation based on matrix "**T**."

Let "**R**" and "**C**" be the sum of rows and columns of the matrix "**T**."

"**R**" and "**C**" is then calculated using eqs. (5.5) and (5.6):

$$R = \left[\sum_{j=1}^{n} t_{ij} \right]_{n\times 1} \tag{5.5}$$

$$C = \left[\sum_{i=1}^{n} t_{ij} \right]_{n\times 1} \tag{5.6}$$

"**R**" represents the sum of influence the element "T_i" executes on others, whereas; "**C**" represents the sum of influence received by T_j from other elements.

"**R** + **C**" is called "prominence," which denotes the degree of influence of one element over the others or the degree by which it is influenced by other elements, whereas "**R** – **C**" is called "relation".

Step 6: Preparation of a causal diagram.

A causal and effect graph is prepared with the horizontal axis representing the "**R** + **C**" values and the vertical axis representing the "**R** – **C**" values.

The following steps present the implementation of the MMDE algorithm for obtaining the threshold value.

Step 7: The $n \times n$ matrix "**T**" is converted into an ordered set T, $(t_{11}, t_{12}, \ldots, t_{21} t_{22}, \ldots, t_{nn})$, to arrange the elements in descending order, thus obtaining ordered triplets T^k in the form of $(t_{ij}\ x_i\ x_j)$.

Step 8: An ordered dispatch node set, T^{Di}, is created considering the second element, that is, dispatch node from the ordered triplets of set T^k.

Step 9: Probabilities are assigned to various elements taking first "t" elements of "T^{Di}" as new set T_t^{Di}, and calculate the H^D of the set T_t^{Di}, H_t^{Di}. Hence, the mean de-entropy value is calculated by $MDE_t^{Di} = \frac{H_t^{Di}}{N(T_t^{Di})}$. The t value is determined by first setting it to be one and then raising it from 1 to $C(T^{Di})$ in increments of 1.

Step 10: The MMDE and T_t^{Di} are picked from $C(T^{Di})$ mean de-entropy values and designated as T_{max}^{Di}.

Step 11: Similarly, T^{re} and $T_{\max}^{Re}$ are derived similarly following steps 7–9.

Step 12: A subset, T^{Th}, is obtained by taking the first "u" elements in T^*. This subset includes all $T_{\max}^{Re}$ elements in the dispatch node and all the $T_{\max}^{Re}$ elements in the receive node. The threshold value is reflected in the minimum influence value in T^{Th} as follows:

$$1 < C\left(T^{Th}\right) < C\left(T^*\right) \tag{5.7}$$

ISM algorithm is implemented after **Step 12.**

Step 13: The total relation matrix obtained after applying DEMATEL is converted into an initial reachability matrix by setting the values inside the matrix to 0 or 1 based on it being lesser than or greater than the threshold value obtained by applying the MMDE technique, respectively.

Step 14: The initial reachability matrix is then checked for transitivity, and the necessary transition links are incorporated.

Step 15: The final reachability matrix obtained is then partitioned into distinct "levels," and a hierarchical structure is prepared in finalized ISM model.

Step 16: Finally, based on the model, a directed graph is drawn.

5.5 Results and discussion

Thus, the tool was tested on the data available for enablers by industry professionals considering the published paper as mentioned above.

The average matrix is available in the paper with the parameters directly inputted into the program.

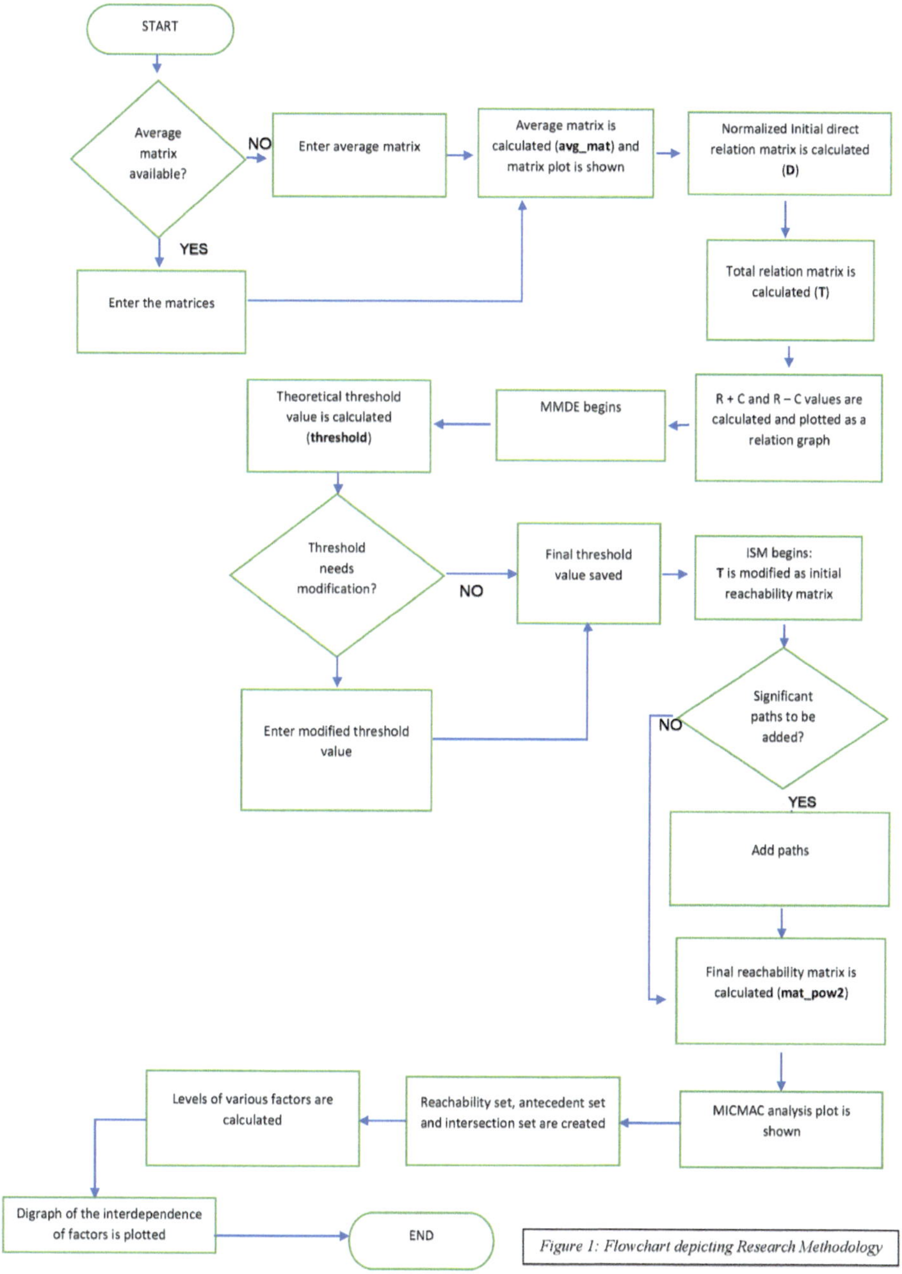

Figure 1: Flowchart depicting research methodology.

Step 1 and Step 2: A single matrix was entered into the program, which was treated as the average matrix comprising of the values of corresponding to the ten enablers. The same matrix was saved as a **Result 1 – Average_Matrix.csv** file, and a matrix plot of the average matrix was created.

Table 1: Average matrix of 10 enablers.

	E1	E2	E3	E4	E5	E6	E7	E8	E9	E10
E1	0	2.2	3	3	3.2	3	2.2	2.2	2.2	2.2
E2	2	0	2.8	2.6	1.8	2.6	2.6	3.2	2.8	2.8
E3	1.8	2	0	2.6	3	3	1.8	2.6	2.4	2.2
E4	2.6	1.6	2	0	2.6	3	1.8	1.2	1	2
E5	3.4	1.6	3.2	1.8	0	1.6	1.2	1	1.2	1.4
E6	2.6	1.4	2.8	2.6	1.8	0	1.8	2.4	1.2	1.8
E7	1.6	1.6	2.2	1.6	2	2.2	0	1	0.8	1.8
E8	2.2	1.8	2.4	2.4	1.8	2.2	1.6	0	2.4	3.2
E9	1.6	2	2.8	1.6	1.8	0.8	0.4	3	0	2.6
E10	2	2.4	2.6	2	2.4	2.2	2.2	2.6	2.2	0

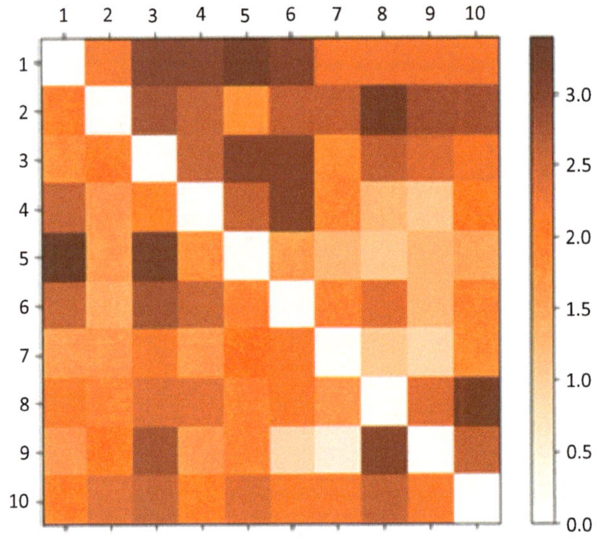

Figure 2: Matrix plot for average matrix.

Step 3 and Step 4: Initial direct relation matrix was obtained, and from that, the total relation matrix was calculated and saved as a **Result 2 – Total_Relation_Matrix.csv** file.

Table 2: Total relation matrix.

	E1	E2	E3	E4	E5	E6	E7	E8	E9	E10
E1	0.436	0.449	0.622	0.555	0.568	0.565	0.431	0.5	0.441	0.514
E2	0.511	0.367	0.615	0.541	0.518	0.551	0.446	0.539	0.466	0.54
E3	0.475	0.415	0.472	0.508	0.526	0.53	0.39	0.484	0.422	0.483
E4	0.444	0.351	0.48	0.35	0.453	0.471	0.345	0.375	0.322	0.414
E5	0.452	0.337	0.501	0.403	0.339	0.404	0.309	0.353	0.318	0.377
E6	0.454	0.355	0.521	0.462	0.439	0.372	0.355	0.431	0.341	0.422
E7	0.353	0.305	0.423	0.357	0.376	0.387	0.232	0.317	0.271	0.354
E8	0.466	0.393	0.54	0.481	0.465	0.482	0.368	0.369	0.408	0.5
E9	0.391	0.356	0.49	0.398	0.409	0.376	0.281	0.432	0.276	0.428
E10	0.469	0.421	0.557	0.476	0.494	0.491	0.397	0.475	0.408	0.389

Step 5 and Step 6: The cause and effect analysis was saved as **Result 6 – Cause_Effect_Analysis.csv** file, and the relation graph was plotted.

Table 3: Cause–effect analysis.

Factors	R	C	R + C	R - C
E1	5.081	4.451	9.532	0.63
E2	5.094	3.749	8.843	1.345
E3	4.705	5.221	9.926	−0.516
E4	4.005	4.531	8.536	−0.526
E5	3.793	4.587	8.38	−0.794
E6	4.152	4.629	8.781	−0.477
E7	3.375	3.554	6.929	−0.179
E8	4.472	4.275	8.747	0.197
E9	3.837	3.673	7.51	0.164
E10	4.577	4.421	8.998	0.156

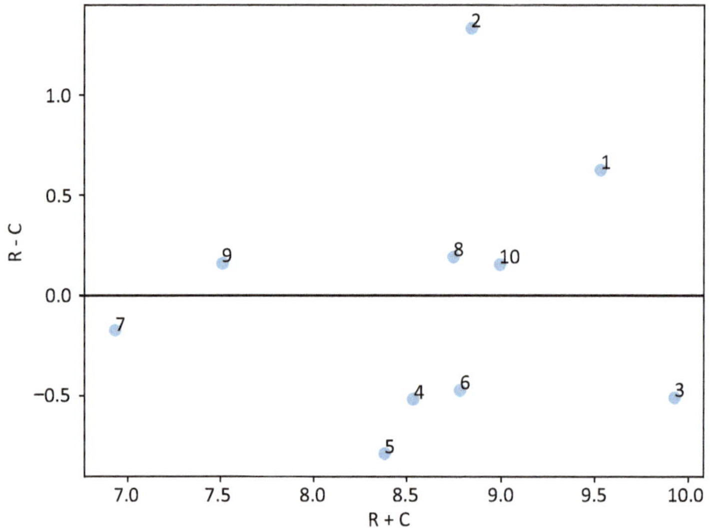

Figure 3: Relation Graph.

Step 7–Step 12: MMDE algorithm was implemented, and a threshold value of 0.557 was calculated. The program prompted to choose to modify the threshold value. No modification was done.

Step 13–Step 16: ISM algorithm implementation was started and the initial reachability matrix was calculated and a few additional paths were added between E2 and E4, E2 and E8, E6 and E5, E7 and E4, E7 and E10, E8 and E3, E9 and E3, and E9 and E10, and the modified initial reachability matrix was saved as **Result 3 – Initial_Reachability_Matrix.csv.**

Table 4: Initial reachability matrix.

	E1	E2	E3	E4	E5	E6	E7	E8	E9	E10
E1	1	0	1	0	1	1	0	0	0	0
E2	0	1	1	1	0	0	0	1	0	0
E3	0	0	1	0	0	0	0	0	0	0
E4	0	0	0	1	0	0	0	0	0	0
E5	0	0	0	0	1	0	0	0	0	0
E6	0	0	0	0	1	1	0	0	0	0
E7	0	0	0	1	0	0	1	0	0	1
E8	0	0	1	0	0	0	0	1	0	0
E9	0	0	1	0	0	0	0	0	1	1
E10	0	0	1	0	0	0	0	0	0	1

The final reachability matrix was calculated and saved as **Result 4 – Final_Reachability_Matrix.csv**

Table 5: Final reachability matrix.

	E1	E2	E3	E4	E5	E6	E7	E8	E9	E10
E1	1	0	1	0	1	1	0	0	0	0
E2	0	1	1	1	0	0	0	1	0	0
E3	0	0	1	0	0	0	0	0	0	0
E4	0	0	0	1	0	0	0	0	0	0
E5	0	0	0	0	1	0	0	0	0	0
E6	0	0	0	0	1	1	0	0	0	0
E7	0	0	1	1	0	0	1	0	0	1
E8	0	0	1	0	0	0	0	1	0	0
E9	0	0	1	0	0	0	0	0	1	1
E10	0	0	1	0	0	0	0	0	0	1

MICMAC analysis was plotted as follows:

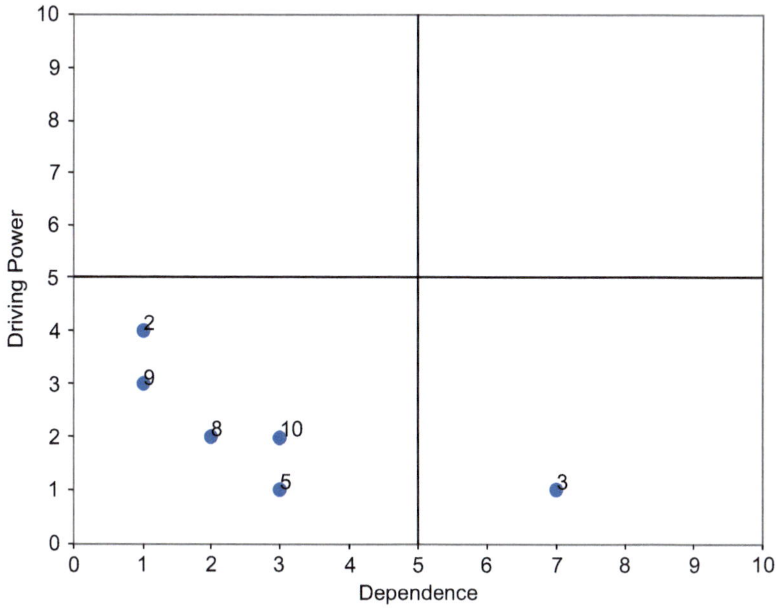

Figure 4: MICMAC.

The reachability, intersection, and antecedent set was calculated and saved as **Result 5 – reach_ante_inter_level.csv**

Table 6: Reachability set, antecedent set, intersection set, and levels.

Factors	Reachability set	Antecedent set	Intersection set	Levels
E1	[1, 3, 5, 6]	[1]	[1]	3
E2	[2, 3, 4, 8]	[2]	[2]	3
E3	[3]	[1, 2, 3, 7, 8, 9, 10]	[3]	1
E4	[4]	[2, 4, 7]	[4]	1
E5	[5]	[1, 5, 6]	[5]	1
E6	[5, 6]	[1, 6]	[6]	2
E7	[3, 4, 7, 10]	[7]	[7]	3
E8	[3, 8]	[2, 8]	[8]	2
E9	[3, 9, 10]	[9]	[9]	3
E10	[3, 10]	[7, 9, 10]	[10]	2

And the final digraph was plotted:

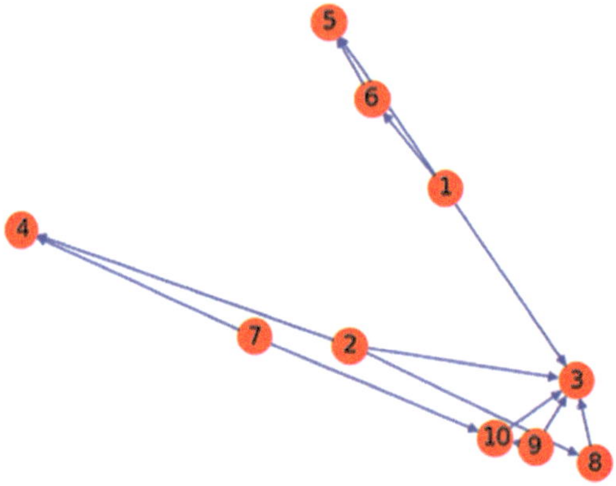

Figure 5: Digraph showing interdependence of enablers.

The Micmac analysis was obtained in the console as follows:

Autonomous factors (lying between dependence < 5 and driving power < 5): [1, 2, 4, 5, 6, 7, 8, 9, 10]

Dependent factors (lying between dependence > 5 and driving power < 5): [3]

Linkage factors (lying between dependence > 5 and driving power > 5): []

Driver factors (lying between dependence < 5 and driving power > 5): []

Miscellaneous factors (lying on the axes): []

5.6 Conclusion

The program runs smoothly and provides a systematic and accurate interface to the user to apply DEMATEL, MMDE, and ISM. The results obtained are accurate, and the time taken is significantly less. This program might prove very useful for management scholars for carrying out cause–effect analysis and tedious calculations. The tool increases the scope of applying these techniques that can be considered for evaluation and are infinitely many without holding back about due to the tiresome calculations.

References

[1] Kumar, A. & Dixit, G. (2018). Evaluating critical barriers to implementation of WEEE management using DEMATEL approach. Resources, Conservation and Recycling, 131, 101–121.
[2] Zhao, Z. Y., Chen, Y. L. & Li, H. (2019). What affects the development of renewable energy power generation projects in China: ISM analysis. Renewable energy, 131, 506–517.
[3] Kapse, C. P., Kumar, A., Dash, M. K., Zavadskas, E. K. & Luthra, S. (2018). Developing textile entrepreneurial inclination model by integrating experts mining and ISM-MICMAC. International Journal of Production Research, 56(14), 4709–4728.
[4] Jain, V. & Raj, T. (2016). Modeling and analysis of FMS performance variables by ISM, SEM and GTMA approach. International journal of production economics, 171, 84–96.
[5] Govindan, K., Palaniappan, M., Zhu, Q. & Kannan, D. (2012). Analysis of third party reverse logistics provider using interpretive structural modeling. International Journal of Production Economics, 140(1), 204–211.
[6] Kamble, S. S., Gunasekaran, A. & Sharma, R. (2018). Analysis of the driving and dependence power of barriers to adopt industry 4.0 in Indian manufacturing industry. Computers in Industry, 101, 107–119.
[7] Mangla, S. K., Luthra, S., Rich, N., Kumar, D., Rana, N. P. & Dwivedi, Y. K. (2018). Enablers to implement sustainable initiatives in agri-food supply chains. International Journal of Production Economics, 203, 379–393.
[8] Zhang, L., Sun, X. & Xue, H. (2019). Identifying critical risks in Sponge City PPP projects using DEMATEL method: A case study of China. Journal of Cleaner Production, 226, 949–958.
[9] Zhou, F., Lim, M. K., He, Y., Lin, Y. & Chen, S. (2019). End-of-life vehicle (ELV) recycling management: Improving performance using an ISM approach. Journal of Cleaner Production, 228, 231–243.
[10] Raj, T., Shankar, R. & An, S. M. (2008). ISM approach for modelling the enablers of flexible manufacturing system: the case for India. International Journal of Production Research, 46(24), 6883–6912.
[11] Panahifar, F., Byrne, P. J. & Heavey, C. (2014). ISM analysis of CPFR implementation barriers. International Journal of Production Research, 52(18), 5255–5272.
[12] Lin, X., Cui, S., Han, Y., Geng, Z. & Zhong, Y. (2019). An improved ISM method based on GRA for hierarchical analyzing the influencing factors of food safety. Food control, 99, 48–56.
[13] Bhanot, N., Rao, P. V. & Deshmukh, S. G. (2017). An integrated approach for analysing the enablers and barriers of sustainable manufacturing. Journal of Cleaner Production, 142(4), 4412–4439.

Shubhika Gaur, Vibha Maheshwari
Risk analysis and management in projects

Abstract: Indeed, according to the saying of masters, evaluating the risk and management is a fusion and obligatory step in the management of project that insures its triumph. Risk analysis and management is considered to be a very crucial part in the field of project management. It is based on the principle, that is, to minimize the cost-effective analysis about the effects and risks involved in the particular project. This chapter explains the various types of project management approaches based on the business strategy. This chapter also highlights the principles and methods that were developed to conceptualize, access, and manage risk. In the field of project management, for the effective and efficient management of the project, in-depth knowledge of the principles and methods of the risk analysis is much needed at every aspect of the project management. Risk assessment and management life cycle has encountered very pivotal importance in today's competitive era. As competition is increasing day by day, this risk analysis can be done in a proper way for the smooth functioning of the business. Although this chapter emphasizes that there are various methodologies available for companies, they have to use them carefully, cleverly, and efficiently during the whole project management.

Keywords: risk management, project management, conceptualize, life cycle

1 Introduction

One of the foremost roles and responsibilities of a project manager is to manage the risk involved in a project on a daily basis and achieve the success of the project effectively and efficiently. An ideal project manager is the one who is able to manage the risk in a systematic manner. Therefore, he/she has to apply various project methodologies and under various types of risk that may be involved in diverse types of projects [1]. Business practices identify, evaluate, track, and mitigate the risks present in the business environment risk management, which play a vital part to devaluate the risk factor. Whether a small-scale business or a large-scale business, the practice of risk management helps in every aspect of risk analysis. Every organization or business wants to achieve stability as they grow in future. For achieving the stability in the organization, the management of risk that affects the business in every aspect has to

Shubhika Gaur, G L Bajaj Institute of Management, Noida, Uttar Pradesh, India, e-mail: Shubhikagaur6@gmail.com
Vibha Maheshwari, G L Bajaj Institute of Management, Noida, Uttar Pradesh, India, e-mail: kabravibha@gmail.com

https://doi.org/10.1515/9783110713633-006

evaluate properly for the success of growth. If the risk aspects of the business are not evaluated in the proper time frame, it will lead to huge losses for the business in the long run. Nowadays, organizations, whether a small or large enterprise, have their dedicated risk management department for the management of the risk. The role of the person who works in risk management domain has to monitor the organization and its environment in every aspect. A business entity can portend the risk factor always by applying the knowledge related to risk management in the advantageous position.

For a competitive advantage than other business, it is crucial to have an eye track on various types of risks involved in the project and accordingly management of risk is important [2]. The main function of the management of the risk is to identify the danger and threats and accordingly manage the risk for the triumph of the project. Management of the risk provides the in-depth knowledge of the various types risk and by this project manager can estimate which type of methodology can be used to improve the efficiency of the project. For efficient allocation, the resources of the organization risk management are very crucial to know each and every aspect. By having the knowledge of the threats of the business, a project manager can also reveal stakeholders about the strength and weakness of the business.

2 What is investigation of risk?

Risk inspection or analysis is defined as the series of undertaking of risk management planning, investigation of risks, and identification and controlling risk on a project for the success of enterprise.

The impact of the risk event through risk analysis helps in examining how project outcomes and objectives might change. Once the risks are understand, they are analyzed to identify the qualitative and quantitative impact of the risk on the project so that appropriate steps can be taken to mitigate them. The following guidelines are used to analyze risks [3].

> Risk management is an integrated process of delineating specific areas of risk, developing a comprehensive plan, integrating the plan, and conducting the ongoing evaluation. –Dr. P.K. Gupta

> Managing the risk can involve taking out insurance against a loss, hedging a loan against interest-rate rises, and protecting an investment against a fall in interest rates. –Oxford Business Dictionary

2.1 Why is risk management important?

For the success of any entity and business, it is very important for the business manager and owner and stakeholder to know about the dangers and threats that are affecting the business. There are several threats that hamper the business, that is, internal threat and external threat. So, for the attainment of any objective and goal, it is needed to understand each and every parameters of the risk involved in different projects and manage them effectively and efficiently.

2.2 In what manner risk can be managed?

Management of risk involves a series of framework of various activities that are crucial to be taken care efficiently [4]. There are seven main stages related to the management of threat in project management. The process of risk management starts by establishing the context and ends with the last step, that is, review and evaluation of the plan. as shown in figure 1.

Figure 1: Risk management process.

1. Organize and scheme risk management: This is the first step in the process of risk management that is organized and scheme the risk management. In this step, various activities involve planning about the project guidelines and also map out the extents of the project. Next part under this step is analyzing the objectives of the stakeholders [5]. This step is the basis of this risk management process which plays a crucial role in evaluating the risk parameters in the project. It also helps in estimating the main agenda of the project. This step helps in managing the procedure to define about the various risks involved in the project.

2. Identification of risks involved in projects: This is the next step and is also a crucial one in the process of risk managementThis step helps in evaluating the various types of risks involved in the project and their adverse effects in future. Various old risks can be documented under this process. The various inputs for identification in the step are as follows:
   - Plan of risk management
   - Statement of the future of project
   - Management plan schedule activity
   - Management plan of human resource
   - Extend standard
   - Cost activity estimation
   - Time duration of activity
   - Documents of project
   - Documents of acquirement
   - Conveying management plan
   - Environmental factor of business
   - Strong suit process of organization
   - Performance of quantitative risk analysis
   - Responses of plan risk
   - Informant risks

If a project manager fails to identify risk at this stage, the organization will suffer a huge loss. This step provides the basis in the process of management of risk. There are various types of risk identification methods Valuation or assessment of risk in project. Ones the risk in valuated in the second step of process of risk management this is the another stage in the process that is valuation of risk in project involve. This step deals with the assessment of nature and ratio of the risk in a particular project. Valuation of a risk factor is simple to measure. After the identification of risks, they need to be assessed to analyze their potential intensity of loss and likelihood of occurrence. In case of the value of a lost building, these attributes could be easy to measure, whereas the same is impossible in the case of possibility of an unlikely event to occur.

Thus, in the process of assessment, it is important to make the best-educated estimates feasible so as to correctly prioritize the execution of the risk management plan. The intrinsic problem in the assessment of risk is to ascertain the rate of occurrence because statistical information of past incidents is unavailable. Additionally, analyzing the extremity of the impacts for the immaterial assets is many times problematic. Valuation of assets is another problem that needs to be considered. Hence, the primary sources of information are best educated estimates (opinions) and available statistics. Notwithstanding, assessment of risk should yield such information for the management of the organization that key risks are understandable and manageable.

Therefore, several theories and attempts have been there to quantify risks. Various risk quantification formulae exist but probably the most commonly accepted formula is the one that is the product of the rate of occurrence and impact of the event [6]. It is crucial to record the findings of risk assessment in financial terms in business. Robert Courtney Jr. (IBM. 1970) propounded a formula to present risks in financial terms. The US government agencies accepted the Courtney formula as the official risk analysis method. The formula presents the computation of annualized loss expectancy and compares the loss value to the security control implementation costs (cost–benefit analysis).

2.3 Potential risk treatments

After the risks have been identified and assessed, all techniques to manage the risks can be categorized as follows:

- **Risk transfer**

Risk transfer means that anticipated party relocates whole or part of the losses resulting from a risk to the other party at a price. Insurance contracts essentially comprise risk transfers. Besides insurance devices, some other techniques are also there to transfer risks.

- **Risk avoidance**

Risk avoidance means not executing an activity that involves risk. Negligence may seem the key to all risks but it also means letting go the potential gains that risk acceptance could have bore. Ceasing a business to neglect risks of loss eliminates the probability of earning gains too.

- **Risk retention**

Risk retention means that the losses occurring because of an exposure of risk shall be borne by the party or the organization [7].

Retention of risk is, in general, a conscious decision for the business organizations devised with the attributes written further.

Self-insurance and captive insurance are two methods of risk retention.

- **Risk control**

Either risk avoidance or controlling losses can help in controlling the risks. Avoidance means that either a certain loss exposure is unacquired or an existing one is deserted. Loss control can be achieved in both ways.

3 Create the plan

Combination of methods to be used for every risk needs to be decided. Documentation of each decision of risk management should be done and acquiescence by the appropriate authority of the management should be taken.

The plan of risk management should suggest appropriate and effective controlling measures to manage the risks. Schedule for control execution and person-in-charge are the requisites of a good risk management plan.

The concept of risk management is ancient but yet is not correctly measured. For instance, antivirus software can be installed to combat the problem of high virus risks in computers.

4 Implementation

All the planned methods need to be followed to combat the impacts of risks.

It is advisable to purchase insurance policies for the risks to be relocated to an insurer, neglect all risks that can be neglected without giving up the firm's goals, minimize others, and acquire the rest [8].

4.1 Review and evaluation of the plan

The existing plan of risk management can never be flawless.

Practice, experience, and actual loss results will facilitate alterations in the plan and contribute information to permit the feasible different decisions to be made in bearing the risks [8].

Periodic improvements in the results of risk analysis and management plans should be done because of the following two key reasons:
1. To evaluate whether the initially opted security controls still hold true and effective.
2. To evaluate the probable risk-level changes in the business environment. For instance, information risks are a good example of the rapidly changing business environment.

4.2 Management of risk mechanism and ways

4.2.1 Quantitative risk analysis

Tools and technique are the static examination of the effect of risk which an organization has to face.

4.2.2 Failure modes and effects analysis

It is for the assessment to know in which situation the process is being wrong. If the failure arises, the measures will be taken [9].

- **Sensitivity analysis:** Here different changes are initiated to show the effect of risk. If the predictions fail, this analysis will show what would happen.
- **A decision tree:** It is the diagrammatic representation of decision, facts, figure, and random events which will show the output or result. The decision tree must be computed with the financial values, which shows that the effects of financial terms must be analyzed for different outcomes.
- **Qualitative risk analysis:** It is also a guide which risks focusing more. Qualitative techniques are of many forms and few are explained further.
- **Red, amber, green (RAG):** It divides risk into three groups; the criteria of division will be quality, time, and likelihood of occurrence. Red risks have greater effect whereas green risk has low or no effect.
- **Risk categorization: It** is the process of categorizing the risk of different groups so that it can be managed easily.
- **Risk urgency assessment:** It is used to decrease the risk which is identified by the RAG status. It mainly focuses on the timing element of risk. It gives priority to the most imminent risks.

  Responses to risks: After getting knowledge about the quantitative and qualitative analyses, different responses are put together to address the risk. The below-mentioned responses can be used alone or as a combination:
  Avoidance: The situation where the risk has been changed. So the risk no longer exists.
  Mitigation: Also known as risk reduction. This action is taken to reduce the risk.
  Transference: In this, the risk is transferred to the third party, for example, an insurer.
  Acceptance: Backup plan to deal with the risk.

4.2.3 Risk monitoring

It is a process of keeping track and watch continuously on the ongoing process of managing risk. Risk can be monitored by gathering information, making strategy, and planning how to overcome it. The organization should consider tools and techniques that will be needed to facilitate monitoring and the frequency necessary to monitor the risk.

4.2.4 Status meeting

Its main purpose is to monitor the success of risk management process. Recent survey states that risk is considered as the main problem in the people's mind.

4.2.5 Risk audit

Audit is generally done to know whether the strategy or plans are working on right direction or not. It can be fully documented for the better understanding of the risk. The objective and findings must be clear for risk audit. It also helps to analyze the process of risk management.

Tools and techniques must be used effectively to get the best result of managing risk. It is the time taken to know the exact tool or technique suitable for your business. One of the great things is that if a tool and technique does not suit your business, you can move on to the other one. Some of the tools and techniques the professionals apply on their business to manage the risk, changes, and issues in their business are as follows:

- **Brainstorming:** Before starting the brainstorming, it is necessary to analyze the effect of risk on your business, for the revision of documentation of project, old data must be checked, and experience that we get by the same kind of projects, articles learned, and organizational process assets is used for the brainstorming. The thing that you are not able to recognize may affect the execution of project. If all the research has been done, then start brainstorming with people who have knowledge about the project. An alternative of this is Delphi technique; in this when a request is sent to experts, they reply unknowingly or the manager of the project can also consult to the team members, stakeholders, and other persons who have sufficient knowledge about the same kind of project.
- **Failure modes and effects analysis:** It is for the assessment to know in which situation the process is being wrong. If the failure arises, measures will be taken.
- **Root cause analysis:** It is another way to find the nature of things. It is a systematic way to realize the basic risks that a project carries. It is the tool which tells that the best management has the responsibility to manage the situations of risks. If any problem arises, then the main cause must be found out. In this, sources are analyzed and not the feature. Yet, for the assessment of risk by knowing the mission of main source of inspection, ask: why this takes place? When it takes place? Where it takes place? When these questions are answered, then make a strategy for not doing the same mistake once more.
- **SWOT:** SWOT stands for strength, weakness, opportunities, and threats. It is one of the tools to recognize risks. For the application of this tool, proceed to phrase. Firstly, know the strength and collect the strong points about the project (this thing is also followed in organizations). Now analyze the weakness about

the project and unknown risk that may arise while working on the project. These analyses consider that the cons and pros are the strengths of the project. Here, the opportunities are like the strength, and threats are the weakness. By analyzing all these things we can make our project more effective and efficient for the organization. While drawing the row and column format of these SWOT analyses in which a table of two rows and two columns is drawn, the upper left corner is named as strength and just opposite to the strength is named as weakness, then bottom left side corner is named as opportunities and just its opposites are named as threats. This table shows that the leftmost column shows the positive side of the project and right side shows the negative things or those that need to be kept in mind while working in a project which may act as a barrier in achieving the mission of project.

- **Risk assessment template for information technology (IT):** This tool is specially designed for IT projects, and other projects can also use it. Therefore, an IT risk assessment template provides a list of risk in digits form for keeping it in sequence, with a way to keep the risk in a manageable situation. Primarily, it gives a place to pile up the risks of the project. It is considered in implementation of a project and follows some of the risks which may become truth. The other feature of risk assessment template for IT, as it has an operating system fixed inside it as a calculator, may reduce the probability of risk which may occur with the multiples that can affect the project of an organization. It is a way that can help the project manager to know the potential harm of the risk with many priorities and their way of responding toward the risk which may arise.

5 Explanation of project

The term "project" means a different endeavor of transient in which a planned objective of the organization includes in the manner of success of the projects and the benefits that can be availed in that particular project. The success of any project can be evaluated in terms of the planned objectives of the project that are able to achieve accordingly in that manner that is designed in the beginning of the project. Efficiency of any project can also be analyzed in the timescale in which a project is complete, and budget and cost incurred in completing the particular project [10]. The following are the various elements of the project in which success of the project is evaluated:

- **Time:** It includes various schedules including planned schedules which indicate that the work will be performed in the planned manner or not.
- **Cost:** This element describes about how much money is required for the functioning of the project for the smooth functioning of a project.
- **Quality:** How will fitness for purpose of the deliverables and management processes are assured?

6 Meaning of management of project

For any organization to achieve the targeted objective, goal, and success, it is needed that the projects should be managed in the proper manner with reliable methods used for the analysis of the risk of the project. The process which leads to the achievement of objective and goals in a specific time period is known as the management of project. One confrontation of the management of the project is to attain the targeted goals in a specific time limit which is specified in the project. In the steps of the project management, it is mentioned in the initial step of management of project, that is, documentation of projects. The main elements that need to be evaluated in the beginning are quality, time, and cost, which are incurred in the project. Another challenge on the management of the project is proper utilization of the distributed resources in the proper input so that the predetermined objectives can be attained in the deadline.

The significance of the project management that it will help produce the final report of the project that how much risk involve in the project as well as it will produce the full project with all elements compile with the requirement of the client which will lead to success of the organization.

Management of a project is the process of different applications in which various skills, methods, and knowledge are required for the achievement of the goal of the organization effectively and efficiently in the specific time frame.

There is a difference between the words "management" and "project management" that management is the continuous process and a lifelong process, but in the project management, the process has an end once the project is delivered in the specific time frame. That is why for the management of the various project skilled is require who has the rich knowledge in the different domains of the project and also have the awareness of the internal and external environment of the business.

For the achievement of the goal of the organization, it is mandatory that the company resources can be properly planned, organized, and controlled in the prescribed manner for the effective achievement of the project. All different resources of the organization and department like finance, marketing, and human recourse should work mutually for the success of the project. Nowadays, project management is required in the field in engineering and construction, and health care and IT, which typically have a complex set of components.

Whether it is construction, healthcare, medical, organization, or education institution, the work of the project manager is the same in every industry or sector. The project manager has to perform the same work, that is, in the beginning, he/she has to collect resources, define the objective of the project, evaluate the risk involved in the project, and finally decide that who will work on the project to achieve the goals. Also manage the quality and efficiency and time frame of the project.

There are various project management methods and approaches used for the evaluation of risk in different projects, but projects also follow different stages for the achievement of the goal of the projects.

- Project launch: This is the first step in the stages of the project in which the role of the project manager is to define the project and various guidelines that are involved in the project.
- Designing of project: In this step, various tasks and records, quality level of the project, and the deadliness of the project will be managed properly.
- Execution: In this step, project manager hires the different expertise in the team who has rich knowledge in the project-related work, and also distributes resources among the team members and specifies about the budget of the project.
- Keep an eye on the project: It is the most crucial stage in the process. Here, the project manager has to track the project, which means he/she has to oversee the risk involved in the project.
- Project windup: This is the last step in the process of a project, that is, project windup. In this step, if the project manager delivered the output to the business and client, then the proper project will be closed.

6.1 Obligation of project leader

The function of the leader of a project is very different. There are various works and responsibilities of the project manager which has to be fulfilled for achieving the goal of the organization in a specific time period [11]. Figure 2 shows the different roles that has to be performed by project manager that is briefing about the project, run the project, build the team, lead the team and also look for the different methods and approaches that need in the project for the minimization of the risk factor in the project.

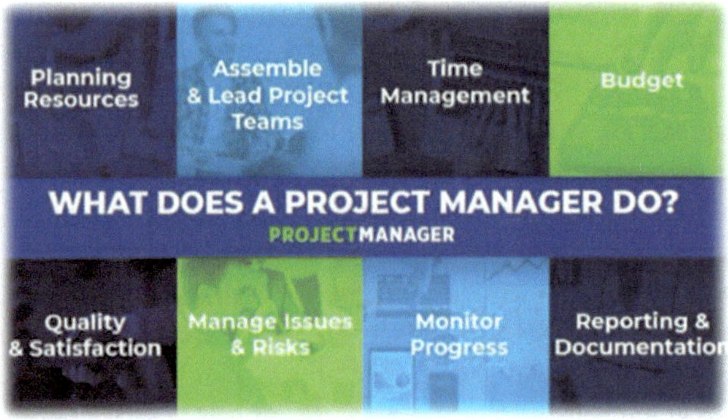

Figure 2: Project manager skills.

6.2 Traits of an efficient and effective project manager

Irrespective of the difference in industries, the work of the project manager is the same in all sectors [12]. The following are the various traits or qualities that a project manager must possess:-

1. Authority skills: By keeping an eye on the project in all aspects, a project manager must possess an authority or leadership skill. For the success of any project in any sector, a strong authority skill is required for efficient achievement of goals of the organization. For a successful completion of the project, a project manager must lead their team from the beginning to the end and must also keep record of the team activities. A good leader is the one who is able to motivate the team for achieving the goals of the organization. A good project manager is the one who is able to understand about the strengths and weaknesses of their team members and to also ensure that the team has all knowledge of the objectives of the project that they need to do in the project as per guidelines.
2. Communication skills: Another important trait that is required for the project manager is the effective communication skill. Communication skill is the essence of all goals and objectives, and without proper communication no activity or task can be carried out properly. A project manager is the mediator between clients, management, and team members so proper communication skill is needed in the traits of the project manager. With efficient communication skill, a project manager will guide the team, motivate the team, and also provide relevant information in relation to the project in the better sense.
3. Analytical skills: In the management of the project, there are various problems that need to be handled and yes for the efficiency of the project each and every problem which comes in the path of the achievement of the goal of the project it is need to be managed by the project manager in the proper manner. A good project manager is the one who is able to resolve all stages of the project and also evaluate and resolve the risks involved in it. It is important for the project manager to address all issues from the management, from client, and also from the team because all play a crucial role in the success of any project.
4. Delegation skills: In the management of the project, assigning the role and responsibilities among the team member as per their area of specialization of their domain is one of the crucial roles of the project manager that need to be addressed properly for the successful achievement of the goal of the organization. In any industry, one trait that is required for the project manager is that he/she should be able to understand the strengths and weaknesses of the team members so that the project manager should be able to delegate the work as per the skills and knowledge of the employees.
5. Team-building skills: Team-building skill is also an important trait that should be required in the project manager. Under team-building skills, it is very crucial that the project leader should know how to motivate and encourage the team

members toward the attainment of the goal. A good and efficient team builder can help in the success of an organization. So, each trait plays its distinct importance in the domain of the project manager.

7 Methodologies of management of project

7.1 Waterfall model

This model was first discovered in 1970 by Dr. Winston Royce for the management of the distinct nature of the software development. This method of project management is one of the oldest methodologies.

It is widely accepted in IT sector that in large due to its efficient and prominent functioning, many software companies adopted this method in project management.

In the waterfall model, a step-by-step representation of various stages is followed as shown in figure 3. Each and every stage, id focus needed. In this model, a project coordinator must have a clear picture of the steps of this model as if one step proceeds further in the next step, and if an error is found in the initial step, then there is no way of any correction in any step in this model. So it is very crucial for the project coordinator that he/she must have proper analysis and evaluation of all stages of this method.

The waterfall model is divided into different crucial steps which begin with the analysis or review of the requirement of the project, collection of the requirement data, analysis of the risk involved in the project, designing of the project implementation technique, after that testing of the project, and ends with the maintenance of the project.

The graphical depiction of the waterfall model is shown in Figure 3.

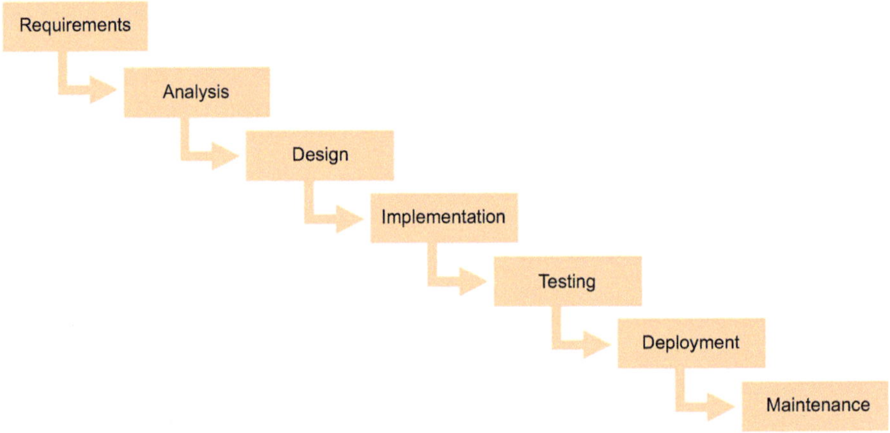

Figure 3: Stages of waterfall model.

Advantages: Various advantages of the waterfall method are as follows:
- Easily understandable: The first advantage of the waterfall method is that it is very easy to understand and use as this model is divided into different stages.
- Design of the waterfall model: The advantage of the waterfall model is that the structure and design of the waterfall model is clear demonstration of the various stages and divided as per the applicability of stages.
- Documentation: The focus of gathering and assessing the understanding heavily depends on the documentation. This helps in new resources which are easy to maintain and move in as per the requirement of the project.

Disadvantages of waterfall model are as follows:
- Risk is more in the waterfall model: As there is no scope of any changes in the stages, if any error found in ant step there is no changes can be made, so the risk factor is highly involved in the waterfall method which leads to the failure of the project.
- Not suitable for the complex projects: Another disadvantage of the waterfall model is that this method is not efficient for the long and complex projects as proper understanding and the knowledge of the projects is required.

The waterfall method is best suited for the below-mentioned project:
- Basic projects
- Project in which all the requirement is clearly defined
- Projects that are based on documentation

7.2 Agile model

This agile method of project management was discovered in 2001. This method was emerged because of the inefficiency of the waterfall model as it is not capable of handling the complex projects. That is why many representatives initiated this agile model to overcome the various issues and problems which came in the management of the project. This agile model is the diverse of the waterfall model. As the name of the agile model explains its working, this model is very fast and flexible. In this method, no gathering requirement is needed.

Figure 4 shows the graphical representation of the agile method.

7.2.1 Advantages

Agile model is fast and flexible: Like the waterfall model, there are no such stages in this method and there is no requirement to focus on this method. This method provides the freedom to implement any incremental changes as per the requirement of the project. This model also promotes the creativity level of the project coordinator also.

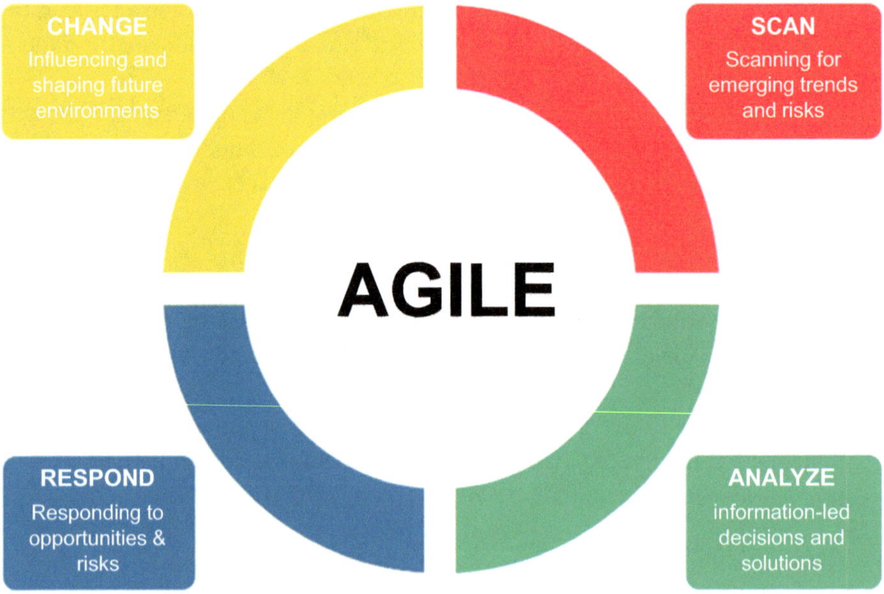

Figure 4: Diagrammatic depiction of agile model.

Risk factor reduces in agile method: While opting for agile method, a project coordinator receives the frequent and fast feedback of the project from the client which will be helpful to maintain the incremental changes in the project.

7.2.2 Disadvantages

- **There is no fixed plan in the agile model: The agile approach emphasizes to respond to changes as they occur. This lack of any fixed plan makes resource management and scheduling harder.**
- **Management of all departments is difficult in the agile model: As in the agile model, there is no fixed plan which directly means that there is a requirement that all departments have to work in collaboration with team. So all the time is not possible that management, stakeholder will always available for the feedback of the changes.**

The agile method is best suited when:
- the project communication and collaboration is not a hurdle but strengthens the project;

- there is no fixed plan; and
- a quick change is accommodated in the project.

7.3 Hybrid

As the name implies about the hybrid model, it is an integrated method which is the combination of waterfall model and agile model. This method is used in diverse types of project. This method is both flexible as well as structured.

7.3.1 Advantages

- **Adjustability is more in hybrid model:** The first advantage of the hybrid model is that it is more flexible than the waterfall and agile models. Because in this model, changes can be requested until the requirement is not substantial.
- **This model is more arranged than the other model:** As this model is the combination of waterfall model and agile model, this model is best for any type of project is taken the planning stage of waterfall model and address the issues of the agile model.

7.3.2 Disadvantages

- **Requires compromise:** As hybrid is the combination of the waterfall and agile models, so compromise will be needed on requirements, and flexibility is required in the model which in many cases is difficult to manage.

The hybrid model is best suited for the projects that are:-
- **in medium size and**
- **with moderate complexity, in which budget is stable.**

7.4 Scrum

Another famous methodology of the project management is the scrum model. Scrum is not a full methodology but it is an extended form of the agile model. This model has the focus on various elements, that are it has a focus on team, sprints, and meetings involved in the project as shown in figure 5. Like agile model, it does not borrow the process. Scrum has its own ways and planning for dealing with the issues in the project management.

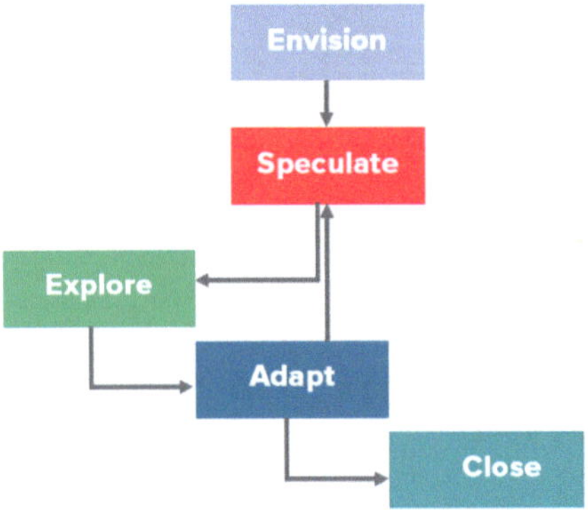

Figure 5: Diagrammatic depiction of scrum model.

The project team has a crucial part in the scrum model. In this model, there is no project manager, that is why the project team is expected to self-organize and manage the project issues themselves.

In the scrum model, there is one requirement, that is, the project team who will handle the project should be highly experienced and knowledgeable.

Agile is the ideology, and Scrum the technique. While scrum is agile, agile isn't scrum.

7.4.1 Advantages

- **"Sprints":** The first advantage of the scrum model is that this model is mainly focused on 30 days "sprints." It is easy for the project team to divide the wish list into small chunks. Then the project team has to work on the project in 30 different sessions with daily meetings which is reliable for the complex and large projects.
- **Fast working in scrum model: For 30 days, this model focuses on sprints and daily meetings which promote the fast working of the model for the complex and large projects and also promote the development in the projects in the long run.**
- **Team-oriented: As in the scrum model, it is required that the project team members should be self-managed and self-focused so this makes this model to be team oriented.**

7.4.2 Disadvantages

- **No fixed deadlines:** As there is a fixed deadline for the completion of project and also no scheduling and budgeting prescribed by the project manager, it can easily lead to scope creep.
- **More threats in the scrum model: Another disadvantage of the scrum model is that as the project team who is working on the project should be**

self-managed so in this model there is higher danger of the failure of this model unless the project team is itself self-motivated by their own. Scrum has a very high chance of failure.

- **Absence of elasticity:** This is the disadvantage of the scrum model, that is, absence of flexibility. This model is not suitable for the large team because it is difficult for the team, and project manager is required for handling the large team.

The scrum model is best suited for the following ones:

- **This is best for the project in which the project team has their own priorities and team should be self-discipline who can manage the project themselves and also aware about the needs of the project clearly.**
- **This model is best suited for the complex and large projects but if a team is large, then this model fails.**

7.5 Critical path method (CPM)

Critical path method (CPM) is one of the popular project methodologies. In this method, first, the activities that need to be completed within the time frame are divided into structures. After the division of the structure duration of single activity, the dependencies structured together for the achievement of the goal. In CPM project, team can first map those activities that are to be completed at the priority basis than those activities that can be mapped out, which has to be done later.

7.5.1 Advantages

- **Efficient arrangement in CPM: The first advantage of the CPM is that it is an effective arrangement of activities, and can be scheduled in an efficient and effective manner without any repetition of the timetable.** For instance, if activity 1 depends on activity 2, CPM will assist and evaluate and schedule those activity which is important first.
- **Prescheduling of an important task: The triumph of the CPM is based on evaluating and lining up the critical and noncritical task. If one's task is prioritized properly as per the requirement, then it can be attained easily.**

7.5.2 Disadvantages

- **Experience requires the prescheduling of the task: The disadvantage of the CPM is that only an experienced project manager who has the real-world**

experience of calculating the task can complete the activity properly, and others cannot handle this model easily.
- **Lack of pliability:** As in waterfall model, CPM is not heavy front, and the project manager has to plan each and every activity from the beginning. In many cases, this model is not suitable for projects where requirements change constantly.

The CPM model is best suited for the project that has independent entity as it works best for those projects. For creative project management, CPM is less suitable.

8 Use of risk evaluation and life cycle evaluation in project management

8.1 What is risk evaluation?

For organizational success, focusing on the evaluation of risk process is very important:
- To find out the causes that impact on people with the help of a process.
- To know problems that would be converted in the severe risk.
- To identify solution taken by the organization to stop problems by controlling severe risk.

To prepare a plan for risk evaluation, the main aim of an organization is to identify the methods for overcome from risk or overall control. Aim includes:
- Giving an evaluation of exact obstacle
- Procuring damages or sicknesses
- Fulfilling legal needs
- Being aware of risk and obstacle
- Making stock-related assets
- Managing the cost of risk should be justified
- Finding out the cost for control and overcome from risks
- Calculating the arrival on savings

Business organizations have various risks, and before commencing a new process in business, they come to know about the procurement of risk and make changes in old processes like changing machinery, or when the corporation detects a new threat. Strong and secure plan of management makes the organization risk free/controlled, thereby adopting a proper process for risk evaluation as shown in figure 6.

8.1.1 Five steps in the risk evaluation procedure

1. Find the threats

   Following is the process for risk evaluation and find out problems for your employees and business it includes:

- Expected tragedies with nature (overflowing, cyclones, storms, quakes, fire, etc.)
- Biotic threats (epidemic illnesses, food-borne illnesses, etc.)
- Work area accidents (slipups and tours, transport accidents, operational failure, mechanical failures, etc.)
- Premeditated acts (laborer strikes, protests, bomb blast, theft, burning, etc.)
- Technical threats (vanished Internet connection, power outage, etc.)
- Chemical threats (asbestos, dusting fluids, etc.)
- Mental threats (surplus workload, harassment, etc.)
- Disruptions in the supply chain

See and observe which type of problems are arising in the organization and focus on the above threats and make a plan and proceed to overcome from these types of situations. You should also focus on all types of happenings and find out what threats have impacted your corporation in the past.

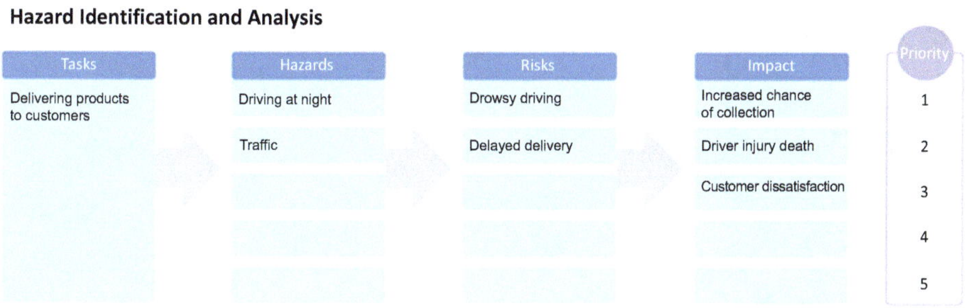

Figure 6: Threat recognition and investigation.

2. Evaluate who might be injured and how

   Focus and identify which threat damages your organization internally and externally, after identifying to find out solutions for controlling such type of problem.

3. To find out the risk and its solution

   There are various techniques and methods to identify risk and then provide a solution. These evaluation techniques help to minimize the risk and prioritize the solution.

4. Keep records of findings

   You can keep records of problems in the organization as per the organizational law based on risk management process and make a plan. This plan includes the

number of problems and procedures to overcome from that situations. Record that plan and explain:
- Direct a suitable check for your workstation to find out who would be impacted
- Overcome and deal with noticeable threats
- Start protections to keep risks short
- Reserve your staff involved in the procedure

5. Analysis of calculation and change if necessary

According to the changes in workplace of your organization, your risk criteria will also change. With new techniques, rules of work, and procedure, problems related to risk will arise and for that regular analysis and changes are required by the organization.

9 Explanation of life cycle assessment (LCA)?

Life cycle assessment (LCA) is the process through which we can extract the raw material and ingredients from environment. Examination of these elements is useful in sustainability of LCA as shown in figure 7. Examination of these elements is useful in sustainability of LCA. LCA is useful in identifying the atmospheric impact of your articles and services from very first to till the end.

Figure 7: Diagrammatic depiction of life cycle assessment.

LCA is useful in many ways. Product designing based on sustainability of the product will impact on product designer for its demand. Policy maker makes plans for promoting, strategic planning, and growth of product. It assists the company for growth. Purchasing section can work accordingly and judge consumer requirement for the product. The best thing about LCA is that it helps in finding the process for assessments of products and business desires in the current situation.

Four steps of LCA

ISO (International Organization for Standardization 14040 and 14044) explains the four main phases of LCA. It provides regularity and consistency:

1. Aim and opportunity definition
2. Stock study
3. Effect examination
4. Explanation

For any investigation, we need more types of data, and LCA helps in improving interactive technology which explains to review your area of opportunity for good results. By learning the top planned LCA, we can recommend valuable suggestions for making various variations in the business.

Step 1. LCA aims opportunities definition

The main aim of LCA experts is to create things simplified and altered, not thinking about the outcome. A model explains about the originality and simplification distorted all around the world. Information is organized very wisely for defining the aims and opportunity of LCA. It is based on product, service, and its scheme of life cycle.

Step 2. Stock analysis of extractions and emissions

The stock analysis explains you about **environmental inputs and outputs** related to goods or service. With the help of an example of an environmental input means thing you take out of the atmosphere to introduce into the product's life cycle which shows "the use of raw things and energy." Ecological outputs explain that product's life cycle puts out into the atmosphere its include the secretion of pollutants and the unused streams. Collectively, this provides you the whole picture.

Step 3. Effect examination (life cycle impact assessment)

Effect examination is the procedure through which we can make better decisions for the business. It estimates that life cycle impact assessment works in the company to interpret as per environmental themes such as global warming and human fitness. The main aim of LCA is to focus on outcome and how much product is sustainable in all types of situation. Choice should be very imperative how collectively they want for better outcome.

Step 4. Interpretation
ISO 14044 standard defines various changes and corrections of test to identify results that data are supported at various uses of actions. At this stage, you can check conclusions are well substantiated. In a similar manner, you can divide your outcomes and enhance conclusions.

9.1 LCA software

LCA software "SimaPro" has been the world's best software since 30 years. This software is useful in LCA for specific search. In more than 80 countries, it is a reliable software according to industry and academics point of view.

With SimaPro, you can:
- know about Key Performance Indicator (KPIs) to evaluate sustainability;
- access your sustainability presentation with life cycle evaluation;
- interconnect clearly the data-based sustainability through result ;
- produce acquiescent ecological product announcement;
- and many more

10 How to choose the accurate procedure

As per the above information, it is better to use Project Management (PM) techniques for various projects. Prism software project would not be used by you, you can work on agile for original growth of estate. We can retain in mind by adopting PM techniques:

10.1 Assess the venture

At the time of selecting the project, we can manage the procedure; it supports for initiating from the edge. Everyone can know about that, how to complete project which we want to see and require exactly after completion.

We can concentrate on the staring need. By using techniques of elasticity, we can able to work on huge and varied team as per the need.

Likewise, we can put organized techniques for exact idea of Waterfall. Iterative policy like agile is useful in indefinite results such as house projects.

We can consider many things for the project as follows:
- Scheme budget
- Timeline
- Size and complication

- Shareholder prospects
- Scheme nature and commerce

10.2 Assess your group

According to your proposal, we can use management techniques. It helps your group what and when to generate. For initiating the project work, your group will come to know every aspects of proposal of that project. If the group has different choices regarding the project, then the group can adopt those choices for getting good results. Group has to focus on learning different techniques to eliminate delays.

In order to organize and assign work to a group, know about the group's strengths and weaknesses. If a group at the time of working on project is flourish and associated, then you can use less organized techniques such as agile but if group is more inspired and self-controlled, a SCRUM aspect can be used. If you have scarcity in usage, then you can use Critical Chain Project Management (CCPM) approach.

Following are some points for analyzing your group:

- Group's knowledge
- Workshops
- Self-organization competences
- Group readiness
- Group's working platform (distant, on-site, etc.)

Essentially, pick a methodology that fits your team, instead of obliging your team to fit the methodology. Here group leader insists the group to adopt one of the techniques which is suitable for them instead of obliging full group to work on the technique.

10.3 Assess your association

Organizational past working and its environment of working also affects on choice of project working and managing different techniques. Few techniques are useful for big organization and making hierarchies but for small organization, learning techniques are more suitable. If we observe that our past and agile project plans have been overdue and not properly established, then we can skip the techniques for future.

Some important points to be considered for organization are:

- Historical data and knowledge with various procedures
- Culture
- Association hierarchy
- Level of elasticity
- Association maturity level
- Association size

- Available possessions, including external resources such as freelancers and servicers
- Your commerce

10.4 Assess your shareholders

When selecting a PM procedure, aspect in:
- Shareholder participation: According to the techniques, shareholder's participation is necessary at every steps. For shareholder's regular response, you can use the agile method. If shareholders are not able to focus on the project, then such type of techniques in that shareholder's involvement is very less.
- Shareholder usages: If the shareholder knows about the project work and what they need for project manager, then they can suggest any type of techniques that are suitable for them, and they can adopt such type of flexible techniques for raising good demand. For more successful proposals, you must know about that shareholders can be happier after keeping all their needs in the mind which is very important.

10.5 Assess your tools

Scheme administration tools are infrequently techniques – agnostic. You can use special techniques for framing a work in a better way. Therefore, you can use software tools for executing and analyzing the impact of your choice of work for that you can do:
- Prepare a list of all software tools that is used regularly.
- Make details of their weaknesses and competences.
- Identify competency of software as per the need of PM techniques.
- You can select techniques for working with the help of software tool instated of buying new software and wasting money and time.
- Working at this path analysis will support for selecting techniques with your targets, your group's competency, and your shareholder impeccably.

11 Conclusion

Project manager works on various project techniques according to their work of proposal. Each technique has its own use, strangeness, and limitations. The project manager should have exact knowledge of the project, and based on that he/she can analyze which techniques are useful for the part of work. All techniques and methodologies are useful for different types of people and work such as stakeholder, industry, software tools, and procedures. Selection of right techniques and methods makes the project more successful. First, we are selecting single techniques from the above list, then we can analyze our project, group, and organization. Shareholders pick their tools software for evaluating strongest and weakness of the project.

According to the study, we can adopt the best methodology for a successful business or project. First, we can observe the strength and weaknesses of our team, organization, shareholder, and availability of software tools and also evaluate which techniques and methods are suitable for them, only then we can adopt the method or techniques for the project.

References

[1] Project Management Institute. (2017). The Standard for Project Management. Newtown Square, PA, Author.

[2] Project Management Institute. (2013). The Standard for Portfolio Management. 3rd edn. Newtown Square, PA, Author.

[3] Project Management Institute. (2017). The Standard for Program Management. Fourth. Newtown Square, PA, Author.

[4] Project Management Institute. (2016). The PMI Lexicon of Project Management Terms. Available from http://www.pmi.org/lexiconterms.

[5] Project Management Institute. Code of Ethics and Professional Conduct. Available from http://www.pmi.org/codeofethics.

[6] Project Management Institute. (2013). Managing Change in Organizations: A Practice Guide. Newtown Square, PA, Author.

[7] Project Management Institute. (2015). Business Analysis for Practitioners: A Practice Guide. Newtown Square, PA, Author.

[8] Project Management Institute. (2014). Implementing Organizational Project Management: A Practice Guide. Newtown Square, PA, Author.

[9] Project Management Institute. 2014. Project Management Institute Excellence in Practice-Research Collaboration, PMI-RI Standards Program: Making Sense of PPP Governance, December 19, 2014. Newtown Square, PA, Author.

[10] Project Management Institute. (2016). Governance of Portfolios, Programs, and Projects: A Practice Guide. Newtown Square, PA, Author.

[11] Project Management Institute. (2013). PMI's Pulse of the Profession® In-Depth Report: The Competitive Advantage of Effective Talent Management. Available from http://www.pmi.org

[12] Project Management Institute. (2015). White Paper, Complexity Management for Projects, Programmes, and Portfolios: An Engineering Systems Perspective, March 2015. Newtown Square, PA, Author.

Sanjive Saxena

Assessing and managing risks in smart computing applications

Abstract: Smart computing applications have *invaded* our lives profoundly. As a result, these applications have become *ubiquitous*. Further, as we continue to fiddle with our *smart device showing* these *smart applications*, there is *always* an *imminent* danger lurking around with the growing usage of this combination. The danger lies in the form of *risks* associated with the exponential growth of consumption of these smart applications, devices, and Internet. However, risks are somewhat unwanted components that are likely to generate a negative impact in the *future*, thus impeding our work and unleashing destruction in one form or the other, if and when it occurs. Hence, it needs to be managed. In other words, it must be planned, assessed, and consequently, mitigation plans be developed to minimize the impact when it occurs.

This chapter deals with the issues of *risk assessment* and *risk management* in the context of smart computing applications. The design of the chapter follows a structured approach. Commencing with the process of identification of *vital* assets and the risks associated with these assets, the chapter then moves to the processes responsible for assessing and managing these risks. The backbone of the chapter is based on two standard models: ISO 27001 and ISO 31000. While ISO 27001 deals with information security management, ISO 31000 deals with risk management.

In today's highly competitive and complex world, the application of knowledge has brought a significant transformation in terms of value addition, growth of the business, and the development of new products and services. This apparent shift toward knowledge engineering, knowledge management, and its application in providing a competitive edge has opened a new arena of information security and risks associated with the mismanagement of information and the consequent impact on business operations. This chapter deals with these aspects.

Finally, as an appendix, the chapter covers a practical application of risk assessment and its management in smart computing applications.

Keywords: data, information, knowledge engineering, smart computing applications

Sanjive Saxena, Jagan Institute of Management Studies, Rohini, Sector, 5, Near Rithala Metro Station, New Delhi 85, India, e-mail: sanjive.saxena@gmail.com

https://doi.org/10.1515/9783110713633-007

1 Introduction

The world today is bombarded by the ubiquitous word *Data*. We all are aware and are familiar with the relevance and importance it has in our life. We have become so dependent on *data* that any form of fiddling, manipulation, tweaking, distortion, misrepresentation, or any other form of associated accouterments results in a feeling of helplessness and despondency. In other words, we dread the consequences of data *mishaps*.

To ensure that these mishaps fail to occur, a systematic approach is needed to identify data, the information that results from processing the data, and the preventive measures that must be undertaken to prevent the occurrence of these so-called data *mishaps*, in other words, issues about risks.

This chapter covers the aspects of risk assessment and risk management in the context of smart computing applications (SCAs).

1.1 Smart computing applications

SCA are the lifeline of today's reality for they have made our life comfortable in innumerable ways. For example, SCA has ensured that we can enjoy our sumptuous food *without venturing out* of our homes at 2 am at night or we can operate our bank accounts without ever going to the bank. While all these amenities are *great*, yet there is a dark side involved. It is our *data* that drives our operations around while we work on our application. For example, booking of train tickets through an *app* on SCA desktop, captures the details of the passengers and ***wow*** it is now available to various stakeholders like banking authorities (*as you have used net banking operations for making payments*) or tax authorities (*as the payment for railway tickets invariably involves some amount of tax*).

To increase the value of SCA, we are required to downloaded an app available in the play store icon on the desktop of smart computing device (SCD). These apps are a bundle of programs developed for a special purpose. For example, the app *Swayam* is developed to ensure that academicians and students log in to upgrade their knowledge, undertake assessment tests, and complete the course [1]. In other words, all the personal information is available on the Swayam server which can *be misused* or *sold*, or *leaked* to a competitor if appropriate measures for the prevention of *data mishaps* are improperly implemented.

With the advent of Internet, several computers are interlinked. Thus, preventive measures for *data mishaps* were easy to implement as the *devices* and *information* were managed *within* and by the business units. However, in the case of SCD, *risks* of *data mishaps* have increased profoundly. The reason is that these devices *are owned* by individuals who *access information about the business operation* by logging into the web browser or downloading the SCA [2].

1.2 Knowledge engineering for modern information systems

The previous section dealt with an overview, relevance, and importance of data in business operations. Further, with the growth of technology, new mechanisms have emerged in terms of accessing data, for example, cloud computing, bring your own device (BYOD), software as a service, and the like. This has large-scale ramifications for security compliance, regulatory compliance, and other forms of measures which, if neglected, may lead to the possibility of happening of *data mishaps*.

1.2.1 Some basic concepts

Before we move further, let us start with data, the basic building blocks of knowledge systems.

Data is the raw fact and figures. The term raw means that on its own it does not have any meaning. In other words, it fails to provide any meaning that may be used in the process of making decisions. Also, data has several forms in terms of structured data, unstructured data, random data, evolutionary data, and the like and *each of these* forms provides an input to the decision-making process.

Information is processed data. As data on its own fails to provide any useful meaning, it must be combined, linked, related, or any other form of processing with another data to generate *some sort of meaningful result* commonly termed as information. It is this information that provides input to the decision-making process. Further, in today's tumultuous business world, effective decision-making must be formalized into a well-developed system that will provide the requisite information by processing structured, unstructured, and other forms of data *needed* to gain a competitive advantage.

To make decisions, information must be available to some reference or context. In the *absence* of context, the information loses its value.

Let us take an example to illustrate the concept of data and information. The display board at the airport depicting the status of flight is information, as the context is to know or determine the exact position or status of the flight so that decision can be taken that is whether to have a cup of coffee or to go home and come to the airport again if one is living close by and the like. This depiction of the status of several flights on the display board may act as data. For example, if the context is to change the destination, due to some emergency, then this may act as data, and accordingly booking may be carried out.

Knowledge is processed information. What it means that when information is combined with another set of information, then we arrive at the knowledge. For example, we may make use of the arrival information of flights and departure information of flights and *process* these two information from different display boards to plan for another travel destination. This is knowledge.

A **system** is an integrated entity comprising people, processes, and tools. People are the stakeholders of the system, while the process defines the constituents of the input, what is done to the input that is the processing part, and what is generated as an output. Worth mentioning is the fact that each of the processes in the system makes use of data to generate the required output. For example, in the railway reservation system, there are processes for reservation of train tickets, cancelation of a train ticket, and the like. Each of these processes makes use of data such as name, age, travel date, and the like. The output is the railway ticket. Tools are the software that is needed to provide a value to stakeholders.

Information system is an organized system that is designed to provide *information* by processing data and communicating this information to stakeholders. To ensure that an information system is effective and efficacious, it operates on the *basic components comprising information* technology infrastructure, processes needed to generate and manage information, and the people needed to manage the information system [3]. Taking the example of a railway reservation system, various types of information are provided by the software such as availability of seats in a particular train, rules, and regulations about changes introduced by railway authorities, and the like. Today, information systems form the core function for various business units due to the fact that in this age of information technology it is the *information* that drives most of the operations carried out in the commercial world.

Knowledge management is a systematic process that utilizes multidisciplinary tools and techniques in the process of identifying, creating, sharing, and managing the knowledge needed to ensure that business objectives are met [4, 5].

Knowledge engineering is the discipline of engineering which is involved in the process of integrating knowledge into a computer-based information system. They are designed to emulate human expertise and knowledge for solving complex problems, in particular, they involve distinct steps for the conversion of human knowledge into knowledge-based systems [6]. The knowledge engineering process includes the following components:

- Knowledge acquisition. It is the process by which knowledge is extracted, converted into a structured format, and is organized in such a manner wherein it can be used in computer software specially developed for solving complex problems in specific verticals or domains. This software is known as expert system.
- Knowledge representation. This is the field of artificial intelligence wherein complex problems of the real world are represented in a form so that they are solved using reasoning.
- Knowledge validation. This is a process that is used to determine whether the knowledge as represented in a system is correct or it is generated with some level of accuracy.
- Inferencing. It is the process of garnering new knowledge that is deriving inference from the existing knowledge. The process of inferencing involves the application of rules and the applicable constraints operating on these rules.

- Explanation and justification. Explanation and justification are the means and mechanism which are used to make the knowledge understandable and not make it understandable but to ensure that it is based on facts. In other words, it is justified and is accepted by the *stakeholder* who is seeking knowledge.

As mentioned earlier, processing of information leads to the generation of knowledge *which* provides the base for optimization or improvement of various processes. Thus, we observe that there exists a hierarchy which is shown in Figure 1.

Figure 1: Depicting stages of data processing resulting in knowledge.

From the above discussion, it is evident that data, information, and knowledge form a crucial element of the decision-making process.

1.2.2 Strategic decision-making and planning in corporate houses

Having understood the basics of data, information, and knowledge engineering, let us now dwell on the role and application of data in corporate houses. Corporate houses need data for carrying out *various operations* such as strategic planning for business expansion, competitor analysis, market expansion, and other forms of crucial operations such as the impact of changes in government policies and the like. Hence, data plays a crucial role. Worth mentioning is the fact that both external and internal data are needed to develop strategic plans for business operations. The process adopted by corporate houses for making decisions starts with *defining* data in the context of determining the *information* requirements, which subsequently will lead to *knowledge* expertise to achieve business objectives such as gain of competitive advantage, optimization of resources, reduction of cost, creation of value, and above all providing assistance in ***dealing with uncertainty***. Hence, a lot of importance is attached in **securing data** and ensuring that adequate measures are identified, implemented, and optimized to ensure that it fails to be exploited by competitors or other entities determined to bring harm to business units [7]. In other words, there is always an element of risk associated with data.

1.2.3 The role of smart computing applications concerning corporate planning

With the rapid and exponential growth of smart computing devices and consequently the generation of a plethora of data, any device can be interconnected and instrumented to achieve the objectives which may be positive or negative. For example, smart cell phones enable an individual to carry on banking activities 24 × 7 × 365 days without venturing ever going to the bank. However, while the foreseen benefits are numerous, dark sides also exist. For example, individuals' cell phones are hacked, and UID's data is hacked from the government server without anyone knowing it [8, 9].

All these events point to one thing that corporate houses are struggling to meet the business objectives. On the one hand, they are required to provide value addition to customers who are now heavily dependent on smart computing devices, but on the other hand, they are equally concerned with the data leakages, security breaches, and the other forms of nefarious activities that motivate the customers to seek value addition to other business units *who at least will provide security of their data* [10, 11].

Corporate planners are now including their *planning* process as to how to manage these smart computing devices to remain trustworthy to their customers. With the onset of cloud computing, IoT (Internet of things), and BYOD gaining proceeded momentum, corporate planners are ardently devising and implementing measures to nullify the threats and vulnerabilities that exist in their risk management process.

Some of the measures corporate is planning to deal with smart computing applications include:

- The concept of the zero trust model. This is the strategy which the corporate is planning as a part of a data security strategy that is designed to meet the complex needs of the present-day work environment. This type of security model is ideal for corporate houses that rely heavily on cloud computing, workforces that operate on remote work environments and distributed systems. The zero trust model also assists the corporate in meeting the compliance requirements of various data security and privacy laws applicable across several countries [12, 13].
- As today's part of the strategic planning process concerning smart computing devices, corporate is laying stress on building up the infrastructure along with the process of planning for operational achievement and excellence. This includes focusing on various aspects such as determining the capabilities of staff members concerning usage, applicability, and skills needed to determine the threats and vulnerabilities that may arise due to smart computing devices [14].

Other measures that are included in the corporate planning process are discussed in the subsequent sections.

Let us look at some of the statistics depicting the risks associated with the data in various business units.

- An average of 5,200 attacks per month are experienced by IoT devices [15].
- High-risk apps are installed on an average of 1 in 36 mobile devices have high-risk apps on their smartphones [15].
- Distributed denial-of-service attacks occurred in February 2018 to the tune of 1.35 terabytes of data, and it is estimated that more of these attacks are likely to occur in the future [16].
- The average cost of a data breach worldwide is to the tune of $3.9 million [17].

- The damages arising due to breach of data on account of cybercrime are estimated to be $6 trillion annually by 2021 [18].

1.2.4 Concept of risk in the context of information systems

Risk, in common parlance, has several concoctions attached to it. For some of the individuals, it means loss of time, loss of money, loss of reputation, and *other forms of losses* depending on the context in question. However, according to ISO 31000:2018, the risk is defined as the effect of *uncertainty* on *objectives* depending on the context [8]. Further, when dealing with risk prevention and mitigation measures, it takes into consideration the following components comprising risk sources, potential events, the consequences associated with these potential events, and the likelihood of the happening of these events.

For example, if we lose our credit card, we face the risk or uncertainty of what will happen next? Will my credit card be used to make heavy purchases? Will my information be misused for other nefarious activities? Or the like. Thus, it is seen that depending on the context, the loss of money or the misuse of information impacts the objectives. Hence, its risk analysis and its management must be *treated with respect*, especially so in this age of information technology where SCA is widely being used to access various information systems *lest* disaster is bound to occur.

The business world of today is highly complex and extremely competitive. To sustain and survive the business environment, business units are required to make decisions, and based on those decisions they need to work out an action plan. Thus, in other words, there is a need to have a structured approach to make decisions as every decision carries with it an element of risk and *it is this risk* that must be accounted for meaning that it must be identified, assessed, and managed so that if that risk were to materialize, it will result in controlling the damages associated with the happening of risk.

The above statistics depict an exigency that measures must be developed to control the damages arising out of risk control measures. In other words, it calls for the development of a structured process for risk identification, risk assessment, and control measures backed by appropriate action plans to control the damages arising out of risk.

1.2.5 Some terms related to risk assessment and management in the context of knowledge engineering

- **Risk management.** It is the process comprising a set of activities all of which co-ordinate with one another to direct and control the business unit in the context of risk [8].

- **Risk assessment.** It is the holistic approach that is designed to identify, analyze, and evaluate the risk. Further, it is a systematic and pro-active approach that involves stakeholders. Being proactive, it is reviewed at periodic intervals, event-driven incidents, and on market dynamics which hinder the achievement of business objectives.
- **Risk identification.** It is the process that is used to determine *events*, *tasks*, *activities*, or *any other* entity which prevents the organization from achieving the business objectives. Worth mentioning is the fact that the driver for risk identification is *information* that must be reliable, accurate, appropriate, and *any other form* of information which will assist the identification of risk.
- **Risk evaluation.** It is the process that involves measuring the outcome of risk analysis with the risk criterion and taking decisions *to determine the course of action plans.*
- **Risk criterion.** It is the process that is used to determine the components of risk that may or may not be taken in the course of pursuing the business objectives. It includes aspects such as the type of risk, the amount of risk, the depth of risk, and the level of significance that must be attached to the process of making decisions.
- **Stakeholder.** It is an entity that is directly or indirectly connected with the risk management program of a business unit.
- **Risk source.** It is the origin or the starting point from which risk can originate.
- **Event.** It is the consequences of a change in circumstances or position that triggers the happening of risk and thereby activating the risk assessment and risk management processes to mitigate the losses.
- **Consequence.** It is the outcome of an event that affects the objectives for which the risk assessment and management activities were put into place. Further, the consequences can be certain or uncertain as they depend on the happening or nonhappening of risk and that these can be expressed quantitatively or qualitatively.
- **Likelihood.** It is the probability of happening or nonhappening of a risk.
- **Control.** It is the mechanism that is designed to measure, maintain, and/or modify the risk.

1.2.6 The process of assessing and managing risks

Having understood the need and importance of risk, let us now dwell on the implementation of risk assessment and management measures.

In general, the following are the basic steps for risk assessment and management:

- Identification of risk objectives that should be in line with the business objectives of the organization
- The management commitment toward risk assessment and management program
- Formulation of risk assessment cell and appointment of chief risk assessment officer

- Identification of assets is crucial for the achievement of business objectives as well as risk objectives along with their access rights
- Formulate information security risk management policies for the identified assets
- Determine the various controls needed for the identified assets
- Formulation and executing of the implementation plan through a process-based approach
- Monitoring of ISMS (information security management system) and conduct of internal audits followed by management review meetings
- Implement continual risk assessment and management plans and policies to keep up with the changing technological advancements

Let us now discuss these points in detail.

1.2.6.1 Identification of risk objectives that should be in line with the business objectives of the organization

This is the most important step in developing the risk management system in an organization. In essence, this means identifying the context of the risk mitigation action plan and which should be in sync with the business objectives of the organization. For example, if the business objective is to increase the turnover of the company in the next 3 years by 15%, then the context of the risk objective would be developed in sync with this business objective. For example, the risk objective could be formulated as to achieve the reduction of development cost by 5%, thereby reducing the risk of not achieving business objective by 20% over last years' risk reduction objective by 15%.

1.2.6.2 The management commitment toward risk assessment and management program

This is the most important and crucial component in the development and implementation of the risk management system in the organization. The absence of management commitment would derail the implementation process *which may impact the achievement of business objectives and in severe cases the survival of the business*. The management commitment should be visible and be demonstrated by every level of the management ladder. An example of management commitment would be the frisking of every individual *irrespective* of the designation. For example, the chief executive officer (CEO) of the organization would be frisked by security guards each time they enter the office premises. This is crucial, as the commitment filters down to the grassroot level. In the same parlance, the chief risk management officer should be subjected to a security check of his/her laptop, desktop, and other accessories carried out by him/her while entering and leaving the office premises.

1.2.6.3 Formulation of risk assessment cell and appointment of chief risk assessment officer

This is the most crucial step. The formulation of a risk assessment cell and the appointment of a chief risk assessment officer ensures that a single point of contact is established in the business unit, and that responsibility and accountability of managing risks lie with the chief risk assessment officer. The chief risk assessment officer would continuously track, monitor, and develop implementation plans, formulation of information security management policies, and discuss the issues and challenges being faced by risk incidents and other forms of events that generate and activate risks.

1.2.6.4 Identification of assets is crucial for the achievement of business objectives as well as risk objectives along with their access rights

Once the risk assessment cell is formulated and the chief risk assessment officer is appointed, the next task which is invariably carried out by the chief risk assessment officer is the identification of assets that are vital for performing the operations of the business. In other words, now the focus of the risk implementation program shifts from macro-level to micro-level. Thus, while the macro-level operations focused on the formulation of risk objectives, risk assessment, and management cell and appointment of chief risk assessment officer, the micro-level operations focused at the granular-level issues such as asset identification, granting of access rights, and assessing the threats and vulnerabilities. In other words, defining the scope of the risk assessment and risk management operations is what all would be covered under the ambit of *risk management*.

1.2.6.5 Formulate information security risk management policies for the identified assets

This is the next step which is most crucial once the crucial assets are identified. In a business unit, each asset has a different priority for the business as well as for the individuals assessing that asset. Hence, policies need to be drafted for identified assets and the individuals accessing that asset. For example, the document containing the financial quote of a particular project is an asset that must be protected at any cost and hence is prone to risk. Thus, this document will not be accessible publicly but it would be available to a limited set of staff members comprising CEO, CFO, and other top management-level executives. On the other hand, HR policies and procedure for taking leave and the leave application form that is needed to be filled in would be made available on the organizational intranet is a public document as it is needed by every employee of the organization.

1.2.6.6 Determine the various controls needed for the identified assets

This is needed to determine the various controls that apply to the identified assets which are risk-prone to the business operations. Not all the assets have the same level of priority in the assessment of risk and its management. While some of the assets have public visibility, others have restricted visibility. This has been discussed in the previous paragraph.

1.2.6.7 Formulation and executing of the implementation plan through a process-based approach

This is the most crucial element of the implementation and sustainability of the risk management program. In essence, this means that there must be an existence of a documented system comprising procedures, policies, checklists, guidelines, and the like. The absence of a documented system possesses the *GREATEST RISK* for the business unit as it is the documented system that mitigates the risk in case of doubts, conflicts, and other forms of inconstancies and ambiguities which are part and parcel of day-to-day operations. The documented system prevents an individual from tweaking the data, the processes, and the procedures which prove to be detrimental to the business operations.

1.2.6.8 Monitoring of ISMS and conduct of internal audits followed by management review meetings

Once the documented system is implemented and *staff members* are trained, the implemented information management systems must be reviewed at defined intervals and event-driven incidences in terms of risks identified by the system, the risk mitigation plans, and other components of an information management system which are centered on the risk assessment and risk mitigation controls.

1.2.6.9 Implement continual risk assessment and management plans and policies to keep up with the changing technological advancements

Once the implementation is complete, it must be monitored for continual improvement. This means that once the system is implemented, the data generated by the system must be analyzed and the processes are required to be tuned accordingly. Also, as the technological changes are happening at tremendous pace, risks associated with technological advancements must be taken into consideration and managed accordingly.

1.2.7 Illustrative example of risk assessment and risk management system in a business unit

ABC is a software development company based in Noida. The company undertakes software development for projects of telecom operators. The company has a staff strength of around 500 and has been in existence since the year 2000. Earlier the company used to develop client–server application software as the technology prevailing at that time was based on the client–server architecture. With the advancement of technology, the company was forced to switch to Web-based software for the customers. This involved a considerable amount of risk. With the passage of time and the advancement of technology and the advent of smart computable devices such as smartphones, the client is not demanding an app-based version of the web-based software developed by the company.

With this background of ABC, over some time, there have been incidents wherein there have been losses to the business, and the risk to the loss is growing day by day. The next paragraph provides a summary of the various incidents that have generated loss to the business as well as it has shown the avenues where possible risks could materialize.

It was the month of December 2009 wherein a group of four laborers walked into the premises of ABC software. The laborers were summoned by the administration department for some repair work on the premises. The work was estimated to be completed in 4 days. As they were required to carry out the repair work, they used to walk in and out of the premises freely *without any* mechanism for controlling. The security guard at the gate never asked their purpose of walking out of the campus nor was involved in frisking their bag and other belonging carried by them. Also, due to the nature of the repair work, they are required to visit one particular area of the premises. After the completion of the repair work, an employee, on the fifth day, reported the matter to the admin department about his missing new notepad which he had requisitioned from the stationary department.

A newly recruited HR executive had a close friend, who was working in the network department and was responsible for maintaining servers and other systems deployed in the organization. The HR executive during lunchtime went to meet his close friend in the server room. They have a good time for 15 min in the server room and later on, it was observed that the Internet had stopped working in one portion of the premises. It was later discovered that the HR executive inadvertently had unplugged the cable as the cable wire got entangled in his feet.

A newly recruited senior executive management of the company was issued a brand new laptop to carry out his work. The executive used to carry his laptop to his home. One evening his son inserted a pen drive in the laptop and switched on the Internet and downloaded his assignment prepared onto his pen drive. The next day, the executive manager observed that his laptop performed erratically.

The above three examples are sufficient enough to depict the possible avenues for risk to various operations being performed for achieving the business objectives.

However, there have been other incidents, more serious and whose impact had been more emphatic were observed at frequent intervals. As the frequency of these incidents started gaining pace, the board of directors of the company and the CEO agreed to adopt a proactive approach to *prevent such occurrences* as these incidents have an impact on the business operations.

It was decided to appoint chief risk officer (*CRO in short*) to address these issues of incidents whose frequency has increased immensely. The CRO was entrusted with the task of implementing a risk management system in the organization. A separate department, named Risk Assessment and Management (*RAM in short*), was created and headed by CRO. To assist CRO in the execution of various tasks, the CRO was provided with a team of eight members. Once the team was formulated, the CRO created the basic risk assessment and risk management structure (Figure 2). This structure was approved by the executive management and was put at various vantage points within the premises to ensure that every employee of ABC understands the importance of risk, the role that they have to play in the prevention and mitigation of risk. Further, by placing this basic structure at several vantage points, it sends the message that management is committed to the attainment of risk objectives while pursuing the business objectives.

Immediately after joining, the CRO commenced his task of formulating his team for the RAM Cell.

The following paragraphs depict how the CRO went about establishing a risk assessment and management system in ABC software.

The first step that CRO took was to identify the areas *which generate information (or data) that are* crucial for the achievement of business objectives. As information is crucial, any *lapse in the management of information* and the related *information systems* has risk potential. In other words, any lapse in *information* generates uncertainty on business objectives.

Further, CRO was well aware that a well-documented information management system must be developed to assess and manage risks in a systematic manner. Further, *policies are the driving force* for assessing and managing risks. Hence, the major work included the formulation of the policies for each of the assets as well as the incidents or events that may happen.

The CRO came up with the following approach and strategy to assess and manage risks.

1.2.7.1 Information security policy document formulation

This is designed to draft means and mechanisms to address the issues about the overall direction of the organization's information security practices. The base for identifying, developing, and forming information security policies of ABC rests on the fact that valuable assets, which generate information *to meet business objectives*, must be protected to provide confidentiality, integrity, and availability of information to *all the stakeholders* who will be using the information in the process of meeting the business objectives. Further, these policies act as the baseline for conducts of audits during periodic intervals, event-driven or incident occurrences.

This process of formulating policies was performed by the CRO and his team and the key holders of ABC in the form of an ***information security management system***

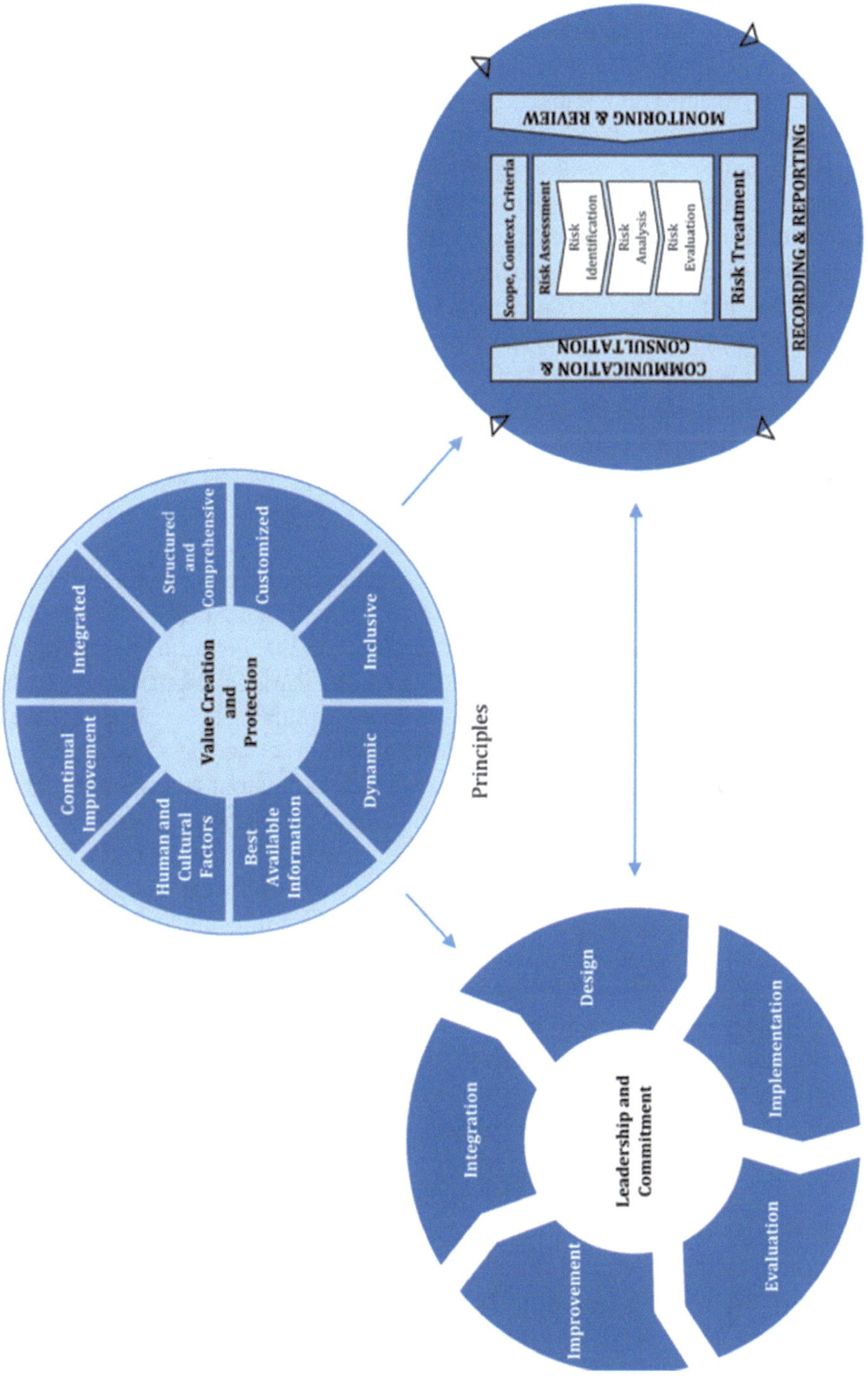

Figure 2: Risk assessment and risk management program structure.
(Source: https://www.iso.org/obp/ui/#iso:std:iso:31000:ed-2:v1:en).

document covering the practical and feasible steps for the implementation of various policies that are applicable from time to time, from event-driven, incident-driven and the prevailing business environment, technological advancements, and the like. The security policies so formulated had the feature of customization wherein a particular or a specific policy can be tweaked to suit the practical implementation and usage in the given circumstances *to meet the business objectives*. However, this should be supported with reasons and approval be obtained. In case, no plausible reason for customization is found, then the employee is submitted to suitable disciplinary action.

The following are some of the key points that were taken into consideration while developing the ISMS document to be applicable across the organization.

- **Scope.** This aspect of the document dealt with the process of fixing the boundaries of the policy. For example, the applicable ISMS policy for working on the laptop *on the premises* is different while the same policy will be tweaked or customized when an employee takes home the laptop. In other words, a separate policy will have the scope worked as *the scope of this policy for working on the laptop is restricted to work within the premises*.
- **Ownership**. This is the crucial aspect of the ISMS policy implementation program and the risk mitigation measures. This puts the onus of the usage of assets onto the owner of the asset concerning confidentiality, availability, and integrity of the information. For example, a stakeholder from the marketing department who is in the process of drafting financials for the project bidding is the owner in terms of confidentiality that the financial details should not be leaked *except* to the authorized persons. In case this is violated and that information is passed on to his junior colleague, the concerned person will be pulled up.
- **Violation**. This is the aspect of the document which dealt with the issue of violations of information security components thereby activating the risk. The essence of this section of the ISMS document lies in the fact of enumerating measures to be adopted in case of violation of information security measures and thereby activating risks. Some of the points of this section included the following aspects:
- Restricting the access rights of the person to assets, the portion of the premises and other areas which carry crucial information and *which may demonstrate breaches to security and thus generate risk*.
- Initiating legal actions in case of breach of information security thereby activating the risk.
- **Security policy statement.** This is the statement that provides the overall direction to the various stakeholders of ABC. This statement clearly states that mechanisms for ensuring that information are secured and the means adopted to prevent the occurrence of risks. The statement formulated by ABC is:

> Information security is the responsibility of every user of ABC information system. Hence the processes of ABC and its policies are formulated to provide the clients with an Information security solution that ensures that the information is integrated, is confidential, is accountable, and

is available whenever needed. Therefore, every employee must be responsible for cultivating a sense of responsibility towards the organizational information through awareness and ownership. A close analysis of the statement indicates that this is prepared by an employee of ABC and sets the direction clear as to what is needed in case of an information breach and the risk associated with it. This security statement is put up at vantage places so that it is visible to several stakeholders. Worth mentioning is the fact that the owner of the security policy statement is CRO *as he is the owner of the information security in the organization.*

- Formulation of teams for handling the various types of risks as a result of information leakage. All these teams report to the CRO.
- **Disaster Prevention Team** is formulated for handling the incidents and carrying out the recovery tasks on the assets which are impacted as a result of the happening of risks.
- **Information Security Operations Team** is formulated for maintaining and overlooking security in normal business operations that take place through ABC's network. In other words, maintain confidentiality and integrity.
- **Information Security Administration Team** is formulated for maintaining security in administrative functions like defining security guidelines, administrative rules, and disciplinary actions.
- **Network and Communications Team is formulated** for providing the proper availability of computing resources to ABC staff members.
- **Technical Support Team** is formulated for providing the necessary technical help to the personnel for carrying out business-related functions.
- **Help Desk Team** is formulated to provide help and redirect distress calls to the respective departments in time of a disaster or an incident.

1.2.7.2 Organization of information security

This aspect of the risk management and risk assessment component takes into consideration the organizational level controls which are needed to ensure that the risk implementation program is established on firm ground. In essence, this aspect takes into consideration by taking into account the process of establishing a framework for information security and that it is regularly monitored, maintained, and updated in the light of the new technological advancements and event-driven. In other words, this aspect of information security management system deals with the formation and working of the system to ensure and assure the security of the flow of information as well as its availability within the premises of the ABC Company. The key aspect of this component is documented in the information security policy document.

The following are key features which the CRO *worked on* and came up with the documented system:

- Commitment to establishing a framework that is responsible for providing clear direction and support to issues on matters of information security. This framework will operate in a structured manner, and the responsibility and authority of

initiating, developing, implementing, and optimizing the framework will rest on CRO who is the management representative also. The CRO was required to choose his team for executing the various tasks about risk mitigation measures that are information security measures in the implementation process. The CRO took into consideration the various aspects of ABC such as business domains, technical domains, administration domains, and other areas including the compliance areas of ISMS based on which risk assessment and risk management action plans will be monitored.
- Defining the responsibility and authority for the implementation of the ISMS framework. This is the most important process for the formulation of any systematic process for implementing it across the organization. To fix the responsibility, the CRO along with the team and in consultation with executive management of ABC formulated the policy that the *owner* that the custodian of the system carries the whole responsibility though he may delegate it to other staff members he cannot shy away from owning the responsibility of the system which carries a risk to the information being leaked or prevents the achievement of business objectives or otherwise. For example, take the case of preparation of a confidential document, say costing of the project for a particular client. The costing process is carried out in consultation with the finance person in addition to the technical head of ABC. In such a case, both the finance person and the technical head have the responsibility concerning the defined scope. Further, the technical head of ABC may delegate responsibility to the project manager. This is a risk as the entire responsibility is on the technical head and if the delegated project manager fails to follow the defined procedures or inadvertently leaks out the information then both the project manager and the technical head will be responsible for activation of risk. They will be subjected to disciplinary actions depending on nature, the extent of the damage, and other implications arising out of the materialization of risk.

Further, to prevent the happening of such types of incidences, the following was designed and developed by the CRO *to ensure that the proposed system to be implemented takes into consideration the maximum points of consideration to minimize the risks*:
- The identification of various assets and their associated processes for security implementation for every individual as well as for every system which helps in the achievement of ABC business objectives as well as risk objectives.
- The owner who is responsible for each of the identified asset or the security process to be followed and their alignment to the documented procedures specifying the steps to be taken at the time of occurrence of the risk.
- Defining and documenting the authorization levels for each of the systems along with the assets involved in the system. In addition to this, the CRO made a provision for management authorization of a new process or a new system of information processing. For example, take the case of a new induction or joining of a new DBA (database

administrator) for a new client having 24 × 7 × 365 online database connectivity. In this case, the CRO must ensure that security policies, procedures, access rights, and background checks for this new DBA have been taken care of by authorized persons. Any lapse or negligence should be accounted for and be disposed of according to the defined procedure. For example, in the recruitment of this DBA, background check was conducted by the HR personnel, but the HR executive did not use the prescribed checklist as prescribed in the information security management framework. As a result, some of the vital points were overlooked which thus formed triggered the risk activation process after 1 month when the DBA had already joined ABC and was handling very confidential aspects of the database. The CRO and his team immediately went into the risk assessment phase and took control of the situation without any significant loss to the client as well as to the business objectives of ABC.

- The CRO focused on developing the processes and policies about checking the hardware and software so that they are compatible with other system components where the threat for risk is extremely high. In addition to this, the CRO developed policies for the usage of personal information processing facilities for business information and for the implementation of necessary controls responsible for risk mitigation strategies. For example, many senior executive management staff members were comfortable using their smartphones and other computing devices; they were asked to provide the details of their devices and an app was installed to prevent them from storing the documents on the local system. Everything was to be done in the cloud.
- Some other aspects which were taken into consideration included the controls for taking the assistance of specialist information security manager wherein some of the threat to information resources is of different nature and the in-house risk management staff members were unaware of tackling the aftermath of the risk happening.
- As the ABC was involved in the process of dealing with third parties for achieving the business objectives, appropriate controls were identified and developed and included the following crucial aspects:
- Formulation, development, and implementation of general policy on information security.
- Permitted access methods and procedures for the stakeholders of a third party.
- Authorized user maintenance about third party usage.
- Respective security, legal, and business liabilities for the stakeholders.
- Audit and monitoring rights and mechanisms for the various operations performed by the parties involved.
- Restrictions on the usage of information copying and disclosure of information that is confidential.
- Antivirus measures to be taken to ensure that it is free from harmful effects of computer antivirus.
- Formulation of arrangements and other measures required for reporting and investigating security incidents and initiation of risk mitigation measures.

1.2.7.3 Asset management

This is the most important and the most crucial part of the risk assessment and risk management implementation program. An asset in a business unit is something that has value to the business unit and which has the potential to assist the management in the achievement of business objectives. It may be a tangible asset or it may be an intangible asset. An example of a tangible asset includes the server, the laptop or radio frequency identification device, and any other device that carries the information which has the potential to be exploited by the competitors. On the other hand, an example of an intangible asset includes the software code which has the potential to be exploited by making minor modifications to the code. Figure 3 provides the matrix depicting some of the matrices and the associated access rights.

Asset	Restricted	Internal	Confidential	Public
Security Policy	Yes	Yes	Yes	No
Changes to Security Policy	Yes	No	No	No
Help Desk	Yes	Yes	No	No
Risk Assessment	Yes	Yes	Yes	No
Information Risk	Yes	Yes	Yes	No
Technical Risk	Yes	Yes	Yes	No
Human Resource Risk	Yes	Yes	Yes	No
Software Risk	Yes**	Yes	No	No
Project Risk	Yes**	Yes	Yes*	No
Third Party Vendor	Yes	Yes	No	No
Server Room	Yes	Yes	Yes	No
Password Management	Yes	No	Yes	No
Access to Cloud	Yes	Yes	Yes	No
BYOD	Yes	Yes	Yes	No
* * Denotes Policy has access to restricted project / persons				

Figure 3: Depiction of various assets of ABC and their access rights.
Source: From author's own creation based on industry experience.

1.2.8 Explanation of access rights

Restricted. It is the classification of information for which unauthorized disclosure or usage may cause damage to the ABC Company, for example, product range design documents and company documentation.

Internal. It is the classification of information that does not require any degree of protection against disclosure within the company, for example, standard operating procedures, policies, and standards interoffice memorandums.

Confidential. It is the classification of information for which unauthorized disclosure or usage is not in the best interest of the organization and/or to its customers, for example, new product design details, strategic planning documents, organization personnel data, and budget information. The loss of such information might be disastrous for the organization.

Public. It is that classification of information that does not require any degree of protection within or outside the company.

1.2.9 Risk assessment

Risk assessment forms the crucial element of the risk management program. The CRO formed took the various points of consideration for formulating the risk assessment methodology. The following are some of the points that were taken into consideration:
- Identification of gaps that exist between the criterion and the current status of the assets which posed threat or which are vulnerable to the occurrence of risks
- Identification of various events or circumstances which may hinder the achievement of the objectives
- Identification of measures that could bridge the gap
- Assessment of risks associated with not implementing the measures for reducing the gap

Figure 4 depicts the risk assessment matrix which the CRO developed for implementing the measures to reduce the risk.

Asset	Source	Threat	Probability	Impact	Vulnerability
Laptop	Unlocked USB Port	Data Leakage / Transfer	Very High	Very High	Can be accessed by Hackers
Laptop	Failure to Format the Laptop before issue	Data Leakage / Transfer	Medium	Medium	Exposure of confidential information to competitors
New Executive	Newspper Advertisement	May not meet expectation as per Job Description	Very High	Medium	Will be a hindrance in meeting team objectives
Sr. Executive	Reference	May leak Confidential information	Very Low	High	May be influenced by previous employer
Chief Finance Officer	Reference	May have attitudinal issues	very high	very high	May be influenced by competitors
Project Manager	Inhouse Promotion	Peers may pose challenge to his style of working	Very Low	Very Low	May be inlfuenced by his achevenment

Figure 4: Risk assessment matrix.
Source: Author's own creation based on his industry experience.

For software projects, the CRO developed a criterion based on the importance, priority, and criticality it has during the development process. Based on these factors, the CRO developed a criticality matrix (Figure 5), based on this criticality score generated by the projects. The criticality matrix formed the ***risk criterion*** from which a decision is needed to be taken.

Engagement Name						
Engagement Code						
Sl. No	Criticality Criterion	Data	Score	Weight in %	Weighted Score	
1	Size of project (Revenue $)	$10,000.00	0.20	10	0.40	
2	Profitability ($)	$40,000.00	1.60	10	3.20	
3	Penalty clause in contract (Amount $)	$50,000.00	2.00	15	6.00	
4	Days left for delivery in working days	1	5	20	20	
5	Resources/ skills availability	Difficulty Level 5	5	10	10	
6	Phase of project	UAT	5	8	8	
7	Type of project	onsite+offshore	9	4	3.6	
8	Impact on client business	High	4	5	5	
9	Link dependency	Link Dependent	1	8	8	
10	IT Infrastructure dependency	Complex	3	5	15	
11	Project Geography	Internal business	2	2	0.8	Recovery Period
			Total	100	83.00	16 Hours
Project Manager						
Date (mm/dd/yyyy)						

Figure 5: Criticality matrix.
Source: Author's own creation based on industry experience.

The criticality matrix depicts the minimum amount of time needed to ensure that the project's resources impacted by the happening of the risk could be made functional in a minimum amount of time. From the figure it is seen that several factors are taken into consideration for ensuring that risk is assessed for each of the ongoing projects, and based on their implementation status, the score is generated along with other components such as financial and impact on client business.

1.2.9.1 Access control

Once the assets were identified, the CRO went on the process of determining the access control of these assets. Each of the identified assets has different access control rights depending on the degree and depth of the usage by the employees of the business units. The CRO came with the following access controls. They are confidential, public, restricted, and private. The example in Figure 3 illustrates the type of access controls depending on the usage by the employee of the business unit and the types of asset that is owned by ABC.

1.2.9.2 Cryptography

Once the access rights are granted to the identified assets, the next step involved the process of allocating cryptographic controls to the identified assets. These controls are identified, designed, and developed to protect assets by ensuring that their confidentiality, integrity, and availability of data are maintained.

1.2.9.3 Physical and environmental security of assets

This is the first point of physical contact of an individual with the business unit. The individual can be an employee, a visitor, or a temporary worker and each of them has a different duration on the premises. For example, a full-time employee has a stay duration of a minimum of 8 h while a visitor may have a stay duration of maybe 1 h or less than that. It all depends on the type and purpose of visit, *and accordingly* control measures are needed to *prevent the leakage of information* and hence risk prevention measures in practice. The longer the stay in the premises, the greater is the threat to information being leaked.

All these factors are taken into consideration while developing measures to assess and manage the risk.

On the environment security, the CRO took into consideration of disposal of outdated electronic equipment such as keyboard, mouse, CPU, storage disk, and any other form of material which once was assessed and is now a liability that it possesses a risk for the business unit in the form of leakage of information *unless* risk mitigation measures are properly implemented.

1.2.9.4 Human resource security

This is the most important component of risk assessment and risk management implementation program *as human resources* are most vulnerable to information leakage and thereby risk to the achievement of business objectives. Hence, the CRO worked on these controls by identifying the past, the present, and at the time of separation of the employee from the business unit.

References

[1] Devi, G. P. & Sornapudi, S. D. (2020). An Insight into digital education in India during COVID-19 from the lens of students. Current Journal of Applied Science and Technology, 39(35), 83–94. doi: https://doi.org/10.9734/cjast/2020/v39i3531057.

[2] Morrow, B. (2012). BYOD security challenges: Control and protect your most sensitive data. Network Security, 2012(12), 5–8, ISSN 1353-4858, https://doi.org/10.1016/S1353-4858(12)70111-3, http://www.sciencedirect.com/science/article/pii/S1353485812701113.
[3] Davis, G. B. (2000). Information Systems Conceptual Foundations: Looking Backward and Forward. In: Baskerville, R., Stage, J. & Degross, J. I. (eds..), Organizational and Social Perspectives on Information Technology. Springer, 61–82.
[4] Girard, J. P. & Girard, J. L. (2015). "Defining knowledge management: Toward an applied compendium" (PDF). Online Journal of Applied Knowledge Management, 3(1), 14.
[5] "Introduction to Knowledge Management". www.unc.edu. University of North Carolina at Chapel Hill. Archived from the original on March 19, 2007. Retrieved 11 September 2014.
[6] Dreyfus, H. L. & Dreyfus, S. E. (1986). Competent systems: The only future for inference-making computers. Future Generation Computer Systems, 2(4), 233–243, ISSN 0167-739X, doi: https://doi.org/10.1016/0167-739X(86)90023-3.
[7] Kavale. (2012). The role of data in strategic decision making process. International journal of current research, 4, 001–007.
[8] Virmani, D., Gupta, C., Bamdev, P. & Jain, P. (2020, Oct 2). iSeePlus: A cost effective smart assistance archetype based on deep learning model for visually impaired. Journal of Information and Optimization Sciences, 41(7), 1741–1756.
[9] Bhadwal, N., Agrawal, P., Madaan, V., Shukla, A. & Kakran, A., Smart Border Surveillance System using Wireless Sensor Network and Computer Vision, International Conference on Automation, Computational and Technology Management (ICACTM'19), 183–190, IEEEXplore.
[10] Agrawal, A., Arora, R., Arora, R. & Agrawal, P., Applications of Artificial Intelligence and Internet of Things for Detection and Future to Fight against COVID-19, A book on Emerging Technologies for battling COVID-19- Applications and Innovations. Springer.
[11] Gupta, C., Agrawal, P., Ahuja, R., Vats, K., Pahuja, C. & Ahuja, T., Pragmatic Analysis of Classification Techniques based on Hyperparameter Tuning for Sentiment Analysis, International Semantic Intelligence Conference (ISIC'21), Delhi, 453–459, 2021.
[12] https://www.iso.org/obp/ui/#iso:std:iso:31000:ed-2:v1:en.
[13] https://phoenixnap.com/blog/data-breach-statistics.
[14] https://www.leadingage.org/case-studies/strategic-planning-and-strategic-it-planning-long-term-and-post-acute-care-ltpac.
[15] https://www.broadcom.com/support/security-center.
[16] https://www.varonis.com/blog/what-is-a-ddos-attack/.
[17] https://www.ibm.com/security/data-breach.
[18] https://cybersecurityventures.com/cybercrime-damages-6-trillion-by-2021/.

Ria Rawal, Kartik Goel, Akshay Gulati, Shivang Sharma,
Palak Girdhar, Charu Gupta, Prateek Agrawal

COVID-19 visualization and exploratory data analysis

Abstract: The outbreak of the novel coronavirus was detected in Wuhan, China, and it has already spread to more than 200 countries now. The novel coronavirus has been named SARS-CoV-2 by the World Health Organization. More than 25 countries had reported their first case of this novel coronavirus and the total number of cases reported worldwide was around 10,000 by the end of January 2020. The WHO declared the outbreak a Public Health Emergency of International Concern on 30 January 2020, and a pandemic on 11 March 2020. There are abundant data available on the Internet regarding coronavirus, including data from official sources such as Government Handles as well as the international organizations such as the WHO and UNICEF. Due to these large amounts of raw data, it is somewhat laborious to dig out the actual facts and grasp the situation of the pandemic. In the wake of this problem, visualization of the raw data helps in providing a clear picture of the pandemic to general masses. This chapter visualizes certain parameters and performs the analysis based on different criteria such as hospitalization and fatality rate which would then be compared across 14 states of the USA. The purpose of this visualization is to help in interpreting patterns so that quick and appropriate actions can be taken in advance which in turn will control the spread of the virus to a greater extent.

Keywords: COVID-19, dynamic modeling and analysis, MERS, Python Library

1 Introduction

Coronaviruses [1] are a large family of viruses that may cause respiration-related illness in humans ranging from common colds to more severe conditions such as severe acute respiratory syndrome (SARS) and Middle Eastern respiratory syndrome. Novel coronavirus is a new [2] and unidentified strain of coronavirus. COVID-19 is the disease caused by this new strain of coronavirus. A series of pneumonia cases [3] were

Ria Rawal, Kartik Goel, Akshay Gulati, Shivang Sharma, Palak Girdhar, Department of Computer Science and Engineering, Bhagwan Parshuram Institute of Technology, Delhi, India
Prateek Agrawal, School of Computer Science Engineering, Lovely Professional University, Punjab, India, e-mail:prateek061186@gmail.com
Charu Gupta, Department of Computer Science and Engineering, Bhagwan Parshuram Institute of Technology, Delhi, India, e-mail: charu.wa1987@gmail.com

https://doi.org/10.1515/9783110713633-008

detected in Wuhan, capital city of Hubei Province in the People's Republic of China. These cases were reported on 31 December 2019. On 7 January 2020, the Chinese health authorities confirmed [4] that this cluster is an outbreak of novel infectious coronavirus. An outbreak is when an illness happens in unexpected high numbers. It may stay in one area or extend more widely. An outbreak can last days or years. A pandemic is a disease outbreak [5] that spreads across countries or continents. It affects more people and takes more lives than an epidemic. The World Health Organization (WHO) [6] declared COVID-19 to be a pandemic when it became clear that the illness was severe and that it was spreading quickly over a wide area.

As more and more cases surfaced, many people and journalists started writing and publishing articles and blogs containing information about this disease [7]. Due to the plethora of data, it was hard to comprehend the ongoing situation of this pandemic, leading to the inception of this chapter to help in visualizing the data in an informative manner [8] so that people can understand the situation. This chapter uses the *New York Times* dataset [9]. They are compiling data from state and local governments and health departments in an attempt to provide a record of the ongoing pandemic. This dataset has been cited by major media companies and agencies such as *CNN*, *Vox*, *The Washington Post*, and *The Wall Street Journal*. The fact that it is cited by such esteemed organizations makes it trustworthy [10] and reliable for general masses. It contains important attributes such as the number of positive cases, negative cases, hospitalized, recovered, and deaths.

The main objective of this chapter is to depict the effect of above-stated attributes in a graphical manner to help assess the gravity of the situation across the states. In this model, COVID-19 Tracker function [11] is created with different parameters such as positive increase, hospitalized increase, death increase, fatality ratio, and total test result increase which are obtained by using the data. Here positive increase signifies the increase in the number of COVID-19 positive cases, whereas hospitalized increase signifies the increase in the number of people admitted to the hospital and death increase signifies the increase in number of deaths due to COVID-19. Furthermore, a comparison of various parameters such as hospitalized increase versus death increase and total test result increase versus positive increase for each state is done across 14 states in the USA. In the proposed model, we adopted methods such as data gathering, data cleaning, data integration for cleaning, and simplifying the data for further analysis. In the process of data gathering, we acquired a series of raw data spread across various columns by the *NY Times* which was further cleaned by eliminating the noise in order to be processed by the model. *Python microfunctions* were used in generating dynamic [12] impactful real-time visualizations. All the functionalities of this model have been explained in detail in Section 4.

The structure of this chapter is as follows: In Section 2, literature review of the referred research papers has been explained, which is followed by Section 3 that is methodology adopted. Furthermore, Section 4 mentions implementation and results, and Section 5 concludes the chapter along with the future scope.

2 Literature review

Worldwide COVID-19 outbreak and the sudden spread of the disease led to large number of COVID-19 cases and deaths. Analysis of the recorded set of data, done on real-time data querying [13], was adopted as a method to gain insights of the disease pattern [14]. Predictive modeling is one such adopted method [15] to interpret the patterns and keep track of the spread of the virulent disease COVID-19. Other patient-related analysis of the data includes physician modeling of patients admitted in ICU and epidemiology model [16]. The epidemic conditions and the economic risks with its major impacts on the world [17] have also been collected in previous papers. However, the collected data analysis is only limited to cases and death. Not only the health risks [18] but fear and panic spread across states and different countries too affect the living conditions of people. This chapter aims to extend the study from previous research works about COVID-19 deaths to parameters such as total patients who are tested for COVID-19, total COVID-19-positive tested people, and total hospitalized people in different US states. COVID-19 fatality rate, that is, the number of people dying from the disease, demonstrated through tabular records and stats has been graphically compared in this chapter.

3 Methodology

In order to analyze the different research aspects of the impact of COVID-19, a set of specific methodologies have been used which covers up different datasets, data sources, modeling techniques, and outcome variables, as shown in Figure 1.

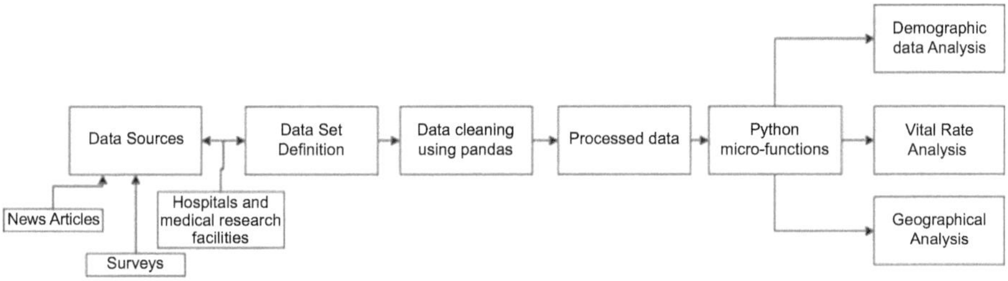

Figure 1: Step-by-step procedure followed for analysis.

The steps of methodology are as follows:

1. Data sources: The basis of any research is the dataset being used. Thus, it begins with the data gathering step in which data has been accumulated from verified sources such as news articles, surveys, hospitals, and medical research facilities.

2. Dataset definition: The verified sources such as news articles, surveys, hospitals, and medical research facilities are cumulated together into a final dataset named as the *NY Times* dataset.
3. Data cleaning using Pandas: The next step deals with data cleaning procedure in which unnecessary or redundant columns have been removed.
4. Processed data: The dataset used is dynamic which means that it gets updated on a regular basis.
5. Python microfunctions: After the processed data is finally gathered, certain functions such as the date function and tracker functions were applied as microfunctions in order to analyze the important parameters of the dataset. These functions help to peek into the dataset and give us many different types of analysis:
   - Demographic analysis answers the question of how COVID-19 has affected the different age groups in a population and which among them has been affected the most.
   - Vital rate analysis answers the question of how many people died due to COVID-19 and help understand the vitality of the population.
   - Geographical analysis helps to throw light upon the comparative analysis between different states in the USA which can further help the researchers to understand the causes and impacts of this pandemic in different regions.

All these different types of analysis together aim to provide an efficacious decision-making system [19], which could help the doctors, epidemiologists, and researchers to fight against thus deadly virus.

4 Experimental results

This chapter analyzes various parameters like positive increase, hospitalized increase, death, total test result, and a comparison among these parameters. The representation of these parameters is given as follows:

4.1 Positive increase

The term positive indicates that the patient is diagnosed with COVID-19. The number of positive cases is the combination of confirmed and probable cases. Probable case is a patient who tests positive via antigen without a positive polymerase chain reaction test.

Figure 2 can be created by using the formula:

Positive increase = number of positive cases (today) – number of positive case (previous day).

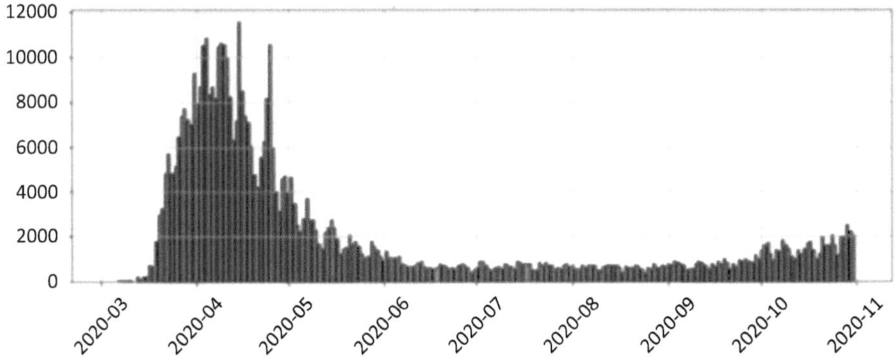

Figure 2: Positive increase for NY.

As observed from Figure 2, the highest peak of positive increase for the state of NY was discovered in the month of April and May in 2020.

4.2 Hospitalized increase

One of the parameters in the dataset is hospitalized increase; this parameter conveys the increment in the hospitalization [20] rate due to COVID-19. Using this parameter, a bar chart is plotted for the state of Georgia (GA) from 2020-03 to 2020-11 (Figure 3).

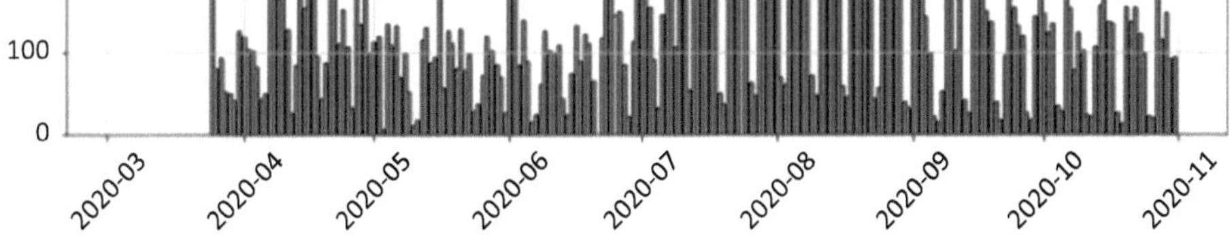

Figure 3: Hospitalized increase for GA.

From this bar chart, we detect a noteworthy increase in the rate of hospitalization in the month of July and August.

4.3 Death increase

After carefully peeking into the dataset, we observe a parameter named as an increase in the number of deaths caused due to COVID-19. A bar chart is plotted for the variable state named Michigan (MI) over a date range from 2020-03 to 2020-11.

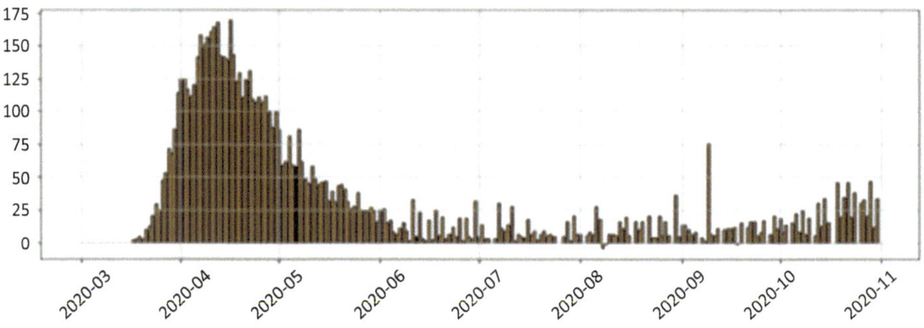

Figure 4: Death increase.

From Figure 4, we can observe that there was a significant increase in the number of deaths between 2020-04 and 2020-05.

4.4 Total test results increase

The graph shown in Figure 5 depicts the relationship between an increase in total test results with respect to the different months of the year 2020. The graph clearly depicts a huge increase in the subsequent months as compared to the initial phase when COVID-19 had just begun to spread in Massachusetts (MA).

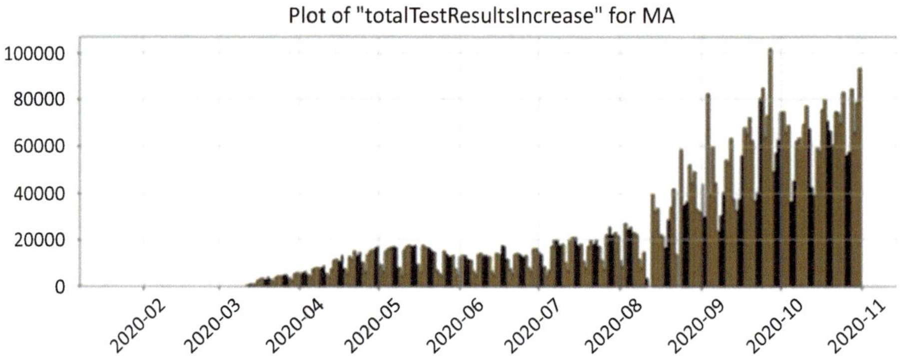

Figure 5: Total test results increase.

4.5 Total test result increase versus positive increase

A graph is plotted between total test result increase versus positive increase for the state of NY, as shown in Figure 6. It depicts the relation between the numbers of COVID-19 positive cases increase with respect to increase in testing. It has been observed that as we increase the number of test, the more asymptomatic cases are diagnosed which result in high number of COVID-19 positive cases but with time more positive cases are isolated which result in decrease of COVID-19 transmission.

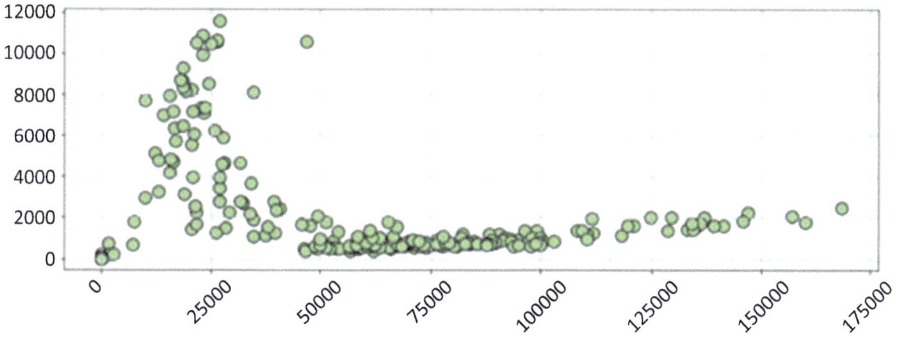

Figure 6: Total test result increase versus Positive increase for NY.

4.6 Hospitalized versus death

A plot function was created to plot the interdependence of two parameters [22] such as hospitalized versus death for the multiple states of the USA (Figure 7). Figure 7 represents the almost linear nature of the two parameters "hospitalized" and "death" for the state of Georgia.

Therefore, such curves help in underscoring the weaker spots in the way of treatment and care given by hospitals in a particular state which could result in the linear nature of the deaths and hospitalized graph.

4.7 Total test results

A testing tracker function was made in order to track the spread of the virus across the various states of the USA. It is one of the most crucial analyses, which can help us control its spread. Figure 8 shows a cumulative testing done by certain states of the USA, which are New York (NY), California (CA), Texas (TX), Pennsylvania (PA), and MA.

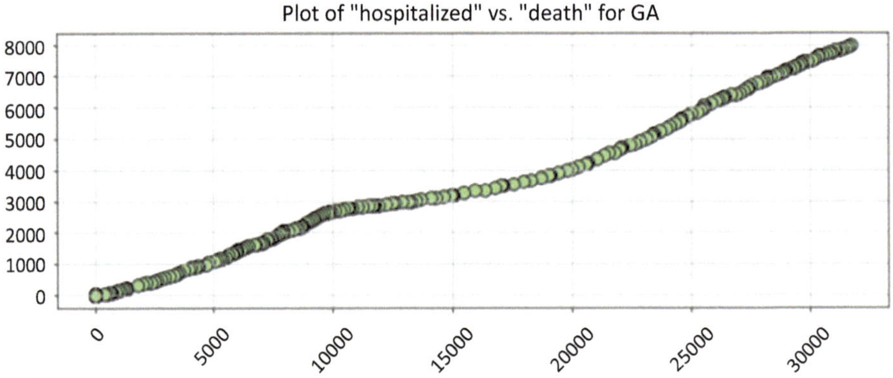

Figure 7: Hospitalized versus death for GA.

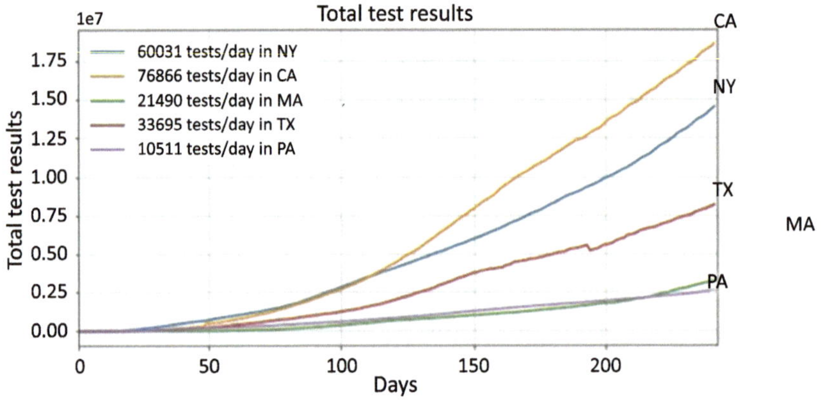

Figure 8: Total test results.

As a result, it was observed that the maximum slope is attained by CA, which in turn depicts that the highest number of tests were being conducted in the state of CA which were approximately about 76,866 tests/day. On the other hand, around 10,511 tests/day are being conducted in PA which are much less than CA.

As per the population survey records, PA has a population around 12.7 million which is around 24.6 million people lesser than CA. It is also around 3.5 times smaller than CA in terms of area. Thus, this difference as shown in the analysis in the total number of tests being conducted highlights the fact that the number of tests being conducted depends highly upon the population of a particular state.

4.8 Fatality ratio chart

Fatality ratio of any region is the clear depiction of the effect, harm, and death the disease causes to the people living in that region. Fatality ratio is calculated by considering the deaths that occurred due to COVID-19 divided by the total number of people who were affected by COVID-19. Here, fatality ratio is calculated for the different states of the USA and is shown in Figure 9.

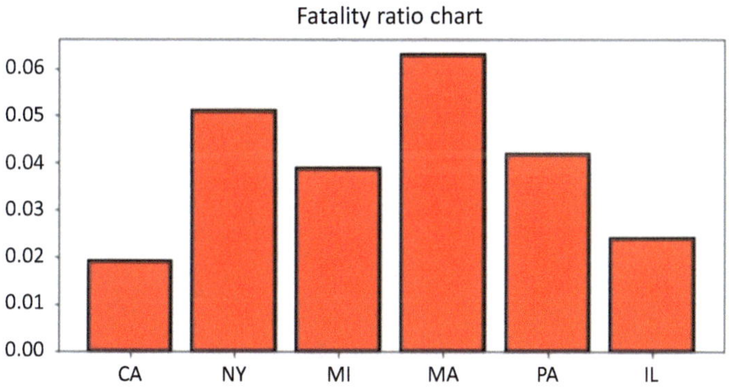

Figure 9: Fatality ratio across different states.

Figure 9 follows the fatality ratio, which is given as follows:

Fatality ratio = Death soccured due to COVID ÷ Total number of people affected by COVID.

From Figure 9, it is clearly visible that MA has the greatest fatality ratio, that is, 0.06, and CA has the lowest fatality ratio, that is, 0.02. This distinctly reflects that the fatality ratio in MA is thrice to that in CA.

On comparing the state of CA with MI, we observe that the fatality ratio in the state of CA is 50% less than that in the state of MI. Furthermore, on comparing the state of MA with MI, we notice that the fatality ratio in the state of MA is approximately 6.7% more than that in the state of MI.

5 Conclusion and future scope

After visualizing different parameters for patients, including the total tested patients, total positive tested patients, total hospitalized patients, total increase in patients, and the fatality ratio of different states, we are able to conclude and summarize the results through *deviation concept* as represented in Figure 10. The results of the difference in new positive cases obtained in the last 14 successive days across CA, GA, Los Angeles (LA) and MI is shown.

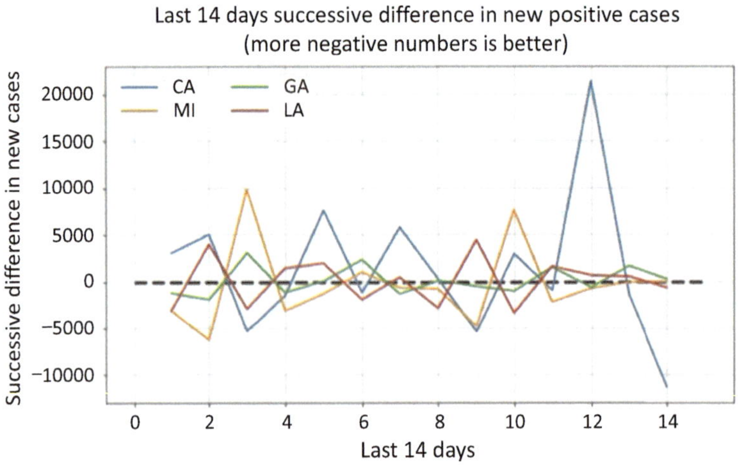

Figure 10: Last 14 days of successive difference in new positive cases.

The graph shows that the maximum deviation in results is seen in the state of CA, where the peak of the difference in the number of positive cases is attained by the end of day 12. On the other hand, the minimum deviation in the difference of the new positive cases is seen in the state of GA, where it is observed to be nearly constant. More the negative numbers the better would be the situation as negative results denote a significant decrease in the number of people afflicted with COVID-19.

The future scope includes extending the analysis by introducing a model which will be based on susceptible, exposed, infectious, and recovered people in population. This basic model could further help in studying the impact of the spread and fatality of the virus, thereby resulting in effective treatment in future.

References

[1] World Health Organization. Novel Coronavirus – China Disease Outbreak News. WHO. 2020. https://www.who.int/csr/don/12-january-2020-novel-coronavirus-china/en/

[2] Glaesser, D., Kester, J., Paulose, H., Alizadeh, A. & Valentin, B. (2017). Global travel patterns: an overview. Journal of Travel Medicine, 24, tax007.

[3] Smith, A. W., Chang, C. R. & Leong, W. Y. (2018). Zika in travellers 1947–2017: a systematic review. Journal of Travel Medicine, 25(1), tay044.

[4] Heywood, A. E. (2018). Measles: a re-emerging problem in migrants and travellers. Journal of Travel Medicine, 25, 1tay118.

[5] Gane, S. B., Kelly, C. & Hopkins, C. Isolated sudden onset anosmia in COVID-19 infection. A novel syndrome? Rhinology.
[6] World Health Organization (WHO), "Coronavirus disease 2019 (COVID-19) Situation Report – 35," WHO, 2020.
[7] Whittington, A. M. et al. Coronavirus: rolling out community testing for COVID-19 in the NHS. BMJ Opinion.
[8] Brann, D. et al. Non-neuronal expression of SARS-CoV-2 entry genes in the olfactory system suggests mechanisms underlying COVID-19-associated anosmia. Preprint at bioRxiv.
[9] NY Times Dataset. https://github.com/tirthajyoti/Covid-19-analysis
[10] Kumaraiah, D., Yip, N., Ivascu, N. & Hill, L. (2020). Innovative ICU physician care models: COVID-19 pandemic at New York-Presbyterian. NEJM Catalyst Innovations in Care Delivery, 1(2).
[11] Real-time tracking of self-reported symptoms to predict potential COVID-19
[12] Gupta, C., Jain, A. & Joshi, N. (2019). DE-For ABSA: A novel approach to forecast automobiles sales using aspect based sentiment analysis and differential evolution. International Journal of Information Retrieval Research (IJIRR), 9(1), 33–49.
[13] Agrawal, A., Arora, R., Arora, R. & Agrawal, P. (2021). Applications of artificial intelligence and internet of things for detection and future directions to fight against COVID. Emerging Technologies for Battling Covid-19, 324, 107.
[14] Rawal, R., Goel, K. & Gupta, C. (2020, November). COVID-19: Disease pattern study based on semantic-web approach using description logic. In 2020 IEEE International Conference for Innovation in Technology (INOCON) (pp. 1–5). IEEE.
[15] Agrawal, P., Madaan, V., Roy, A., Kumari, R. & Deore, H. (2021). FOCOMO: Forecasting and monitoring the worldwide spread of COVID-19 using machine learning methods. Journal of Interdisciplinary Mathematics, 24(2), 443–466.
[16] Holmdahl, I. & Buckee, C. (2020). Wrong but useful – what covid-19 epidemiologic models can and cannot tell us. New England Journal of Medicine, 383(4), 303–305.
[17] Bloom, D. E., Cadarette, D. & Sevilla, J. P. (2018). New and resurgent infectious diseases can have far-reaching economic repercussions. Finance and Development, 55(2), 46–49.
[18] Pfefferbaum, B. & North, C. S. (2020). Mental health and the Covid-19 pandemic. New England Journal of Medicine, 383(6), 510–512.
[19] Madaan, V., Roy, A., Gupta, C., Agrawal, P., Sharma, A., Bologa, C. & Prodan, R. (2021). XCOVNet: Chest X-ray image classification for COVID-19 early detection using convolutional neural networks. New Generation Computing, 1–15.
[20] Omer, S. B., Malani, P. & Del Rio, C. (2020). The COVID-19 pandemic in the US: a clinical update. Jama, 323(18), 1767–1768.
[21] Cepaluni, G., Dorsch, M. & Branyiczki, R. (2020). Political regimes and deaths in the early stages of the COVID-19 pandemic. Available at SSRN 3586767.
[22] Tatar, M., Habibdoust, A. & Wilson, F. A. (2021). Analysis of excess deaths during the COVID-19 pandemic in the state of Florida. American journal of public health, 111(4), 704–707.

A. Ilmudeen

Business intelligence and decision support systems: business applications in the modern information system era

Abstract: Today, the business intelligence and decision support systems are proven as a vital infrastructure for the ever-growing business organization. Business enterprises are ever more depending on data to respond to key operational and strategic operation of their customers, markets, and their stakeholders. The business intelligence has progressed as the volume of data created by the intelligent devices and the Internet have grown up exponentially. Nevertheless, without appropriate applications and systems in a position to analyze the increasing big data, the enterprises are encountering various complexities. This chapter discusses the applications, challenges, and conceptually designed architecture that includes skills requirement, mining techniques, technical, design elements in business intelligence, and decision support systems.

Keywords: business intelligence, decision support systems, business applications, architecture, modern information era.

1 Introduction to business intelligence and decision support systems

The significance of business intelligence and decision support systems is evitable, and it has long been recognized in the business setting. The applications of business intelligence have gained much attention among both the academia and the practitioners. Scholars believe that the business intelligence is one of the foremost significant information technology (IT) applications in business firms and is anticipated to endure so for some time (e.g. [1]). The business intelligence offers dominant approach to help the data-driven decision-making process, which is the central of "business intelligence" [2]. The applications of business intelligence and decision support systems are aiming to reduce the accompanying cost and increase the efficiency of the business operation.

The connection between the business intelligence and the decision support system is inseparable. Accordingly, the business intelligence has turned out to be a complement for the decision support system in enterprises [3]. Some scholars define the business intelligence as the wide-ranging decision support systems in the business organizations [1]. The business intelligence has diverse purposes based on the

https://doi.org/10.1515/9783110713633-009

field; in general, it is a data-driven decision support system that integrates data collection, data storing, and an analysis to enable a decision-making process [4]. However, there is an open debate for the ability of decision support systems that are based on the traditional database and knowledge bases as they are limited because of the lack of information. Furthermore, today, the huge amount of data about the product, customers, business operation, and transaction are stored. However, the main question is how to convert such big data into significant information in order to optimize their profit and revenue, which become a serious concern for businesses [5].

Business intelligence refers to the knowledge, practices, and the applications for collecting, assimilating, analyzing, and the generation of business information. It integrates business analytics, data mining, data tools, knowledge management, data visualization, warehouses, OLAP, infrastructure, and best practices that help firms to make more data-driven decisions. These tools and techniques are not only providing the ability to get required information but also to turn them into suitable intelligence bases that will increase an enterprises' competitive position. The business intelligence includes various technologies, practices, applications, and systems which are used to handle and analyze data to help the business enterprises to understand their market, customers, and business opportunities and well-timed business decisions.

Scholars define the decision support system in numerous ways. For instance, Yun et al. [6] define it as the "combination of the information system and decision-making technology." It is a kind of information systems that helps humans for the complex decision-making processes [7]. The decision support systems enable executives and professionals to understand what information is required, when it is required, and in what form it is needed in order to make smart business decisions. Usually, the decision support systems are used to plan, manage, and support the operational levels of business activities wherein this involves ranking, sorting, and suggesting the best among alternative business options.

Scholars and practitioners agreed that the business intelligence and analytics to expose valuable information to business decision-makers across various levels of an organization to make superior, and more meaningful decisions [8]. Therefore, today the business intelligence systems and data-driven decision support systems have moved the businesses toward more intelligent enterprises. Accordingly, the business intelligence enhances the enterprise's efficacy in handling information for decision making that integrates people, technology, and processes within the enterprise. The business intelligence integrates the data storage and knowledge management that enhances the business decision-making process more efficient.

Figure 1 shows the trend from 2005 to 2020 of the terms "business intelligence" and "decision support system" under the data management category. It shows comparatively the decision support system that marginally received popularity than the business intelligence. But both are progressing similarly in a level over time in data management category. It is obvious that the technological, architectural, platform, and application domains had already been applied in the industries a long years ago.

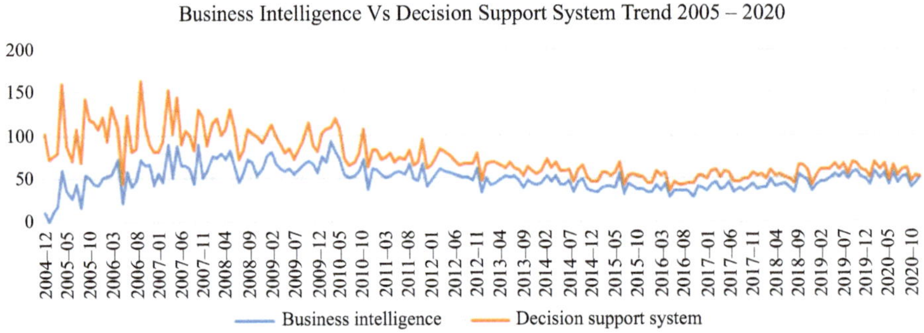

Figure 1: Business intelligence and decision support system trend (source: Google Trends).

This chapter focuses on to review the applications and challenges, and to discuss the conceptually developed architecture in business intelligence and decision support systems. This conceptually designed architecture of this business intelligence and decision support system illustrates the technical, architectural, and domain requirements and features in the modern information system era.

1.1 Applications of business intelligence and decision support systems

In literature, the business intelligence has been applied in various fields for diverse stakeholders' information requirements to take superior decisions [4]; for instance, in managing power and energy [9], hospital management [10], telecommunication sector [11], and tourism sector [5]. Likewise, the utilization of business intelligence healthcare analytics tools improved the decision-making process for healthcare professionals, as a result exploiting the value of clinical data [10]. The business intelligence has developed, as the amount of data made available over the Internet and smart devices has grown-up changing how enterprises use information.

Business intelligence systems are intended to support the firm's knowledge worker communities including directors, experts, analyst, and the managers in the operational level [3]. Currently, businesses are capitalizing through business intelligence and analytics to increase their competitive value additions [8]. The power of business intelligence is to handle data from various business units such as marketing, human resources, accounting, finance, sales, and research and development that in turn encourages to achieve competitive position in the market [12]. The business intelligence and analytics can enable many opportunities from big data, and field-centric analytics are required in some application areas [13].

Business intelligence applications enable easier the enterprises to handle large volume of data, visualization from the data, intelligent analytics, and dashboard reporting. Enterprises can set up business intelligence to systematize the traditional report generation to computerized platforms and assimilate with other enterprise-level systems. Business intelligence systems explore to guess the market demand products and exact revenue sources, hence, the businesses can pay prompt consideration to the unexpected financial crisis and can resolve it quickly. In the marketing domain, the business intelligence systems automate the reporting generation process; thus, it helps the marketing professionals to systemize their personalized, region, and country-based marketing promotions.

Similarly, various researches which discuss the applications of decision support systems in today's business world are manifold; for instance, logistics [14], agribusiness decision [15], clinical and disease analysis [16, 34], higher education [17], enterprise modelling [18], traffic management [7], environment disaster management [19], and smart city development [20]. The decision support systems' modeling and ranking are used to choose the supreme logistics service supplier so as to increase the efficiency in the logistics operation [14].

In the context of agriculture, decision support system can be employed for planning agricultural mission, food waste control, climate change decision, water resources management, and analyzing meteorological, market demand, and land usage data for superior decision making [15]. The web-based decision support system helps to schedule and optimize the resources and time using integer programming in the higher education context [17]. The decision support system is designed to automatically interpret and contextualize the emergency crisis situation to alert and response their users [21]. Internet of things (IoT) and fuzzy rule classification approach are used to design and develop the decision support system to diagnose, predict, and analyze the dengue fever [34]. The decision support systems can assist to project the sales revenues and forecast the inventory levels, production scheduling, and sales comparative analysis for particular time intervals. Table 1 provide the selected studies for the applications of decision support systems and business intelligence in the business domain.

Table 1: Selected studies for the applications of decision support systems and business intelligence in the business domain.

Author(s)	Purpose/problem	Method, technique/ platform features	Application domain	Highlight/findings
Sarabi and Darestani [14]	The decision support system to select the best logistic service provider	The fuzzy best-worst method and multiple objective optimizations based on the ratio analysis and full multiplicative form	Logistic	Among the three efficient logistic service providers selected

Table 1 (continued)

Author(s)	Purpose/problem	Method, technique/ platform features	Application domain	Highlight/findings
Fertier et al. [21]	Decision support system to model emergency crisis situations in real time	Model-driven and event-driven architectures	Contextualization and interpretation of crisis situation	Facilitating emergency managers during the crisis and complex situation
Leung et al. [22]	Re-engineering logistic process of e-order in distribution centers	Using genetic algorithm and rule-based inference and decision support in cloud-based environment	e-Commerce and logistics	Efficient order handling by reducing time and traveling distance for e-orders
Pérez-Castillo et al. [18]	Decision support system modeling the enterprise architecture	Genetic algorithm and mining techniques	Modeling the enterprise architecture	This approach maximized the accuracy and control the cost
Dunkel et al. [7]	Decision support system for sensor-based traffic control management	Event-driven architectures to handle the complex real-time traffic data	Road traffic management	This solution will support to design the intelligent traffic management system
Yun et al. [6]	Visual decision-making system for industrial applications	Data mining technique and decision support system architecture	Data mining	Proposed method proved the effective and strong result compared to other methods
Ilmudeen [23]	Designing IoT-enabled decision support system to predict and analyze dengue	Cloud, fog, and IoT components are integrated and fuzzy rule based-neural classification method applied	Healthcare – dengue analysis and prediction	The design, approach, and method proposed to design and develop this system
Business intelligence studies				
Francia et al. [24]	An augmented business intelligence that offers a set of appropriate analytical queries to the user	Mapping objects, and multidimensional cubes and queries	Real-world devices data and augmented reality	The experiments prove the efficiency, effectiveness, and user satisfaction of the proposed method

Table 1 (continued)

Author(s)	Purpose/problem	Method, technique/ platform features	Application domain	Highlight/findings
Radenković et al. [9]	Designing data flow, forecasting, data analysis, and decision making for the electricity market	Modern business intelligence solution and methods are used	Electricity market	It shows the adaptable and active electricity market management for the data-rich smart grid environments
Niño et al. [12]	Designing of business intelligence governance framework for a university	A detailed diagnosis was carried out to pinpoint the status of maturity	Higher educational institution	The design confirms the achievement of business intelligence projects and alignment of their plan with institution vision
Kao et al. [10]	Design deployment of business intelligence system for hospital	Design science research method used by coveing six activities	Hospital industry	System successfully developed, and implementation shows significant performance improvement
Vajirakachorn and Chongwatpol [5]	Implementation of business intelligence to handle data into meaningful information	Design concepts of database, analytics, performance measurement, and visualization	Tourism sector	Proposed business intelligence framework showed soundness

1.2 Conceptual, methodological, and technical bases of business intelligence and decision support systems

The business intelligence from the IT perspective integrates data and storage with analytical power for the complex and viable knowledge planners and business decision makers. Business intelligence supports to examine the causality and corporate analyses as it deals with a data-driven method to link enterprise strategic priorities and functioning actions [10]. The enterprise architecture has referred to a crucial architectural model of an enterprise's behavior and assets that lead to produce timely business decision [18]. The business intelligence creates significant information to facilitate business decision making [2]. Table 2 shows the methods and techniques in business intelligent and analytics-related domain.

Table 2: Underlying methods and techniques in business intelligent and analytics-related domain.

Data analytics	Big data analytics	Web analytics	Text analytics	Network analytics	Mobile analytics
Time series analysis	Data mining	Information retrieval	Information retrieval	Bibliometric analysis	Web services
Regression analysis	Forecasting and time series models	Computational linguistics	Document representation	Citation network analysis	Smartphone platforms
Predictive modeling	Extract, transform, load	Search engines	Query processing	Co-authorship network	Social media analytics
Pattern recognition	Online analytical processing	Web crawling	Relevance feedback	Social network theories	
Visualization	Regression	Website ranking	Text retrieval, extraction, summarization	Network metrics and topology	
Spatial analysis	Classification and clustering	Search log analysis	Text categorization and clustering	Mathematical network models	
Sentiment analysis	Predictive modeling	Recommender systems	Enterprise search systems	Network visualization	
Simulation	Anomaly detection	Web services			
Social network analysis	Association analysis	Mashups			
Supervised learning	Optimization	Content analysis			
Unsupervised learning	Neural networks				
Statistics	Genetic algorithms				
Signal processing	Multivariate statistical analysis				
Decision trees	Heuristic search				
Fuzzy clustering	Sequential pattern mining				

1.2.1 Applications of modern techniques and methods integrated with business intelligence and decision support systems in the business domain

With its expansion, business intelligence can explore the unseen associations, forecast future consequences, and allocate resources [3]. In the integration of artificial intelligence with IT, the state-of-the-art cloud computing will facilitate the logistics industry practitioners to advance their in-house order handing more efficiently [22]. The following section discusses the modern techniques and methods that are associated with business intelligence and decision support systems in the business domain. Figure 2 shows the different stages and foundations for the evolution of business intelligence.

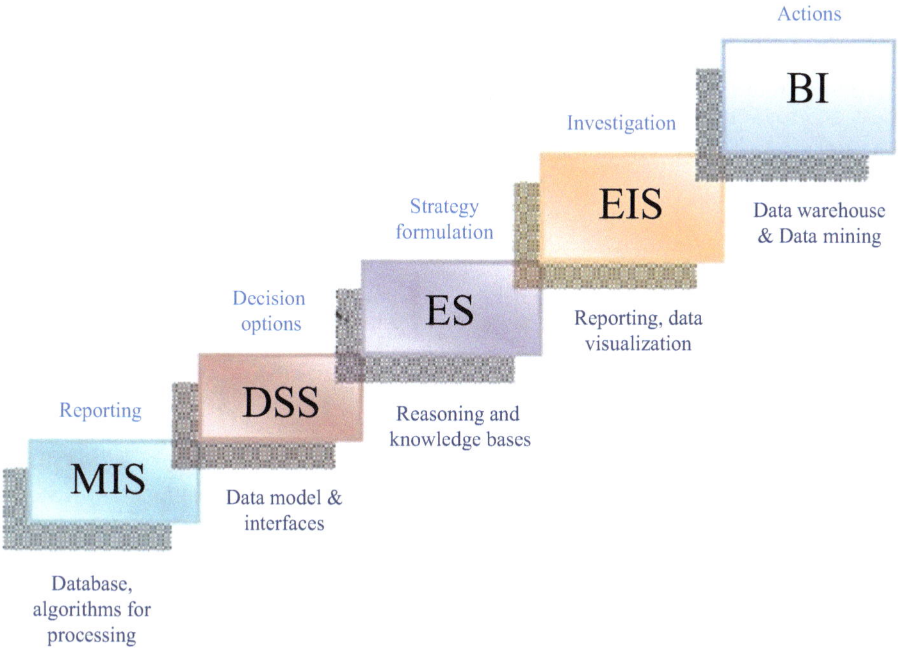

Figure 2: Business intelligence evolution.

1.2.2 Big data

In today's big data era is the acquiring, processing, and diffusion of evidence and knowledge to advance an enterprise's position to enrich its products and services [25]. With the expansion of recent state-of-the-art technologies, the big data can be used for exploring, mining, extracting, visualizing, and simulating hidden insights in a meaningful manner. The enterprises have realized the value of incorporating the modern analytic techniques in recent years, focusing on big data and trying to integrate machine learning and artificial intelligence techniques for superior data analysis. The usages of big data analytics with cloud computing techniques have made possible to handle large set of data [26]. In crosswise businesses, the big data has combined traditional structured data as a mission-critical component. The giant web-based firms such as Facebook, Google, and Amazon are exploiting many opportunities from big data by using advanced machine learning and data mining techniques [27].

1.2.3 Analytics

In today's informative business era, app, smart device, and software platforms either for private or business usages are more and more being developed to store data. With the recent rapid development in the analytics and big data, the usages of business intelligence have been widespread across business enterprises. Until recently, business enterprises are highly centering on predictive and prescriptive exploration that employ machine learning along with rapid analytics over visualization [28]. Now the data analysis plays a vital role in creating knowledge, producing significant trend and forecasting for framing strategies [12]. Scholars agreed that the resilient decision-making culture might considerably influence on crafting a competitive advantage with the power of analytics [8].

Today the power of analytics can explore remarkable prospects in the business. Hence, the business managers leverage from analytics across various functional areas of the enterprise. The recent technologies such as artificial intelligence, cloud computing, and IoT have empowered the analytics more dominant. Analytics offers hidden insights that help business leaders to choose the finest course of action. Businesses must have analytics that permit them to explore their data meaningfully and use it more effectively they have never assumed earlier. Table 2 lists the different types of methods and techniques that are used in the business intelligence and various analytics domain.

1.2.4 Data science, machine learning, and deep learning

The data science is an umbrella term in which the machine learning comes within it, whereas deep learning is a kind of approach in machine learning. Business intelligence is one of the key areas that data science specifically focuses on. The machine learning with its superior analyzing power enables for exploring the unseen trend and hidden insight in the business data. Data scientists have been applying the machine learning models for long time to resolve many difficult and specified business domain problems.

In the old-style dashboard, business managers just observe their sales figure, hence, cannot explore further idea or decision. Yet, there are unseen patterns and trends in the causal arrangement of the sales figures that might bring numerous concern for better decision making. For instance, some products' sales performance is good, whereas others may be poor. This significant intuition is unseen for normal view. With the power of machine learning and deep learning, the automation process helps to find unseen insights that can offer much faster information to act rapidly. Accordingly, the automation process will help the analysts' daily task, such as deviation analysis, finding inconsistencies, and approving clarification in the reports, easier.

1.2.5 Data mining for business intelligence

The data mining is defined as the way to extract required knowledge and information by exploring data from the database, data warehouse, and other sources (POS, logistics supervision tools, and IoT-enabled production equipment, etc.) [6]. Further, the data mining is an essential element of business intelligence, where the enterprises can apply data mining to extract the facts and figures they require with the help of business intelligence and analytics. Data mining technique mines hidden, indefinite knowledge, and hidden pattern that has the potential value for business decision-making. Today, the tendency, for instance data science and analytics, has arisen as the part of business intelligence [28].

The business intelligence can support by integrated services such as online diagnostic handling, business reporting, position intelligence, and dashboard updating. In addition, the business intelligence provides businesses the chance to link heterogeneous databases into one single source, gather and assemble data, and facilitate the workers to mine insights and dashboards in which more up-to-date business decisions can be taken. With the power of analytics, the data and business knowledge can enable to understand the business environment more broadly [8]. The data mining can be applied to extract, analyze, and transform with the various methods to dataset such as clustering, querying, classification, pattern identification, regression, rule-based association, outer recognition, and prediction.

Nowadays, the business intelligence and data mining have supported for various business domains in which the better business decisions can be taken. Further, it is expected that the data mining methods would bring superior business applications that would support for business problems. The data mining techniques, for instance, online analytical processing and advanced querying, help to handle big data that are stored in the data warehouse and data mart to generate more informative business reports in a timely manner. Here, the online analytical processing and data mining tools (e.g., for summarizing data that can further extract, roll in, classify, segment, and cube) are used for reporting and projection [11].

1.2.6 Artificial intelligence

To help machines to recognize patterns and inferences, artificial intelligence experts depend on deep learning and natural language processing [29, 30]. The artificial intelligence is laying its existence all over the place in today's modern information era. The artificial intelligence–powered technologies have noticeable position for improving the competencies of business analytics and intelligence. Artificial intelligence in business is growing as enterprises can optimize machines' algorithms to pinpoint trends and insights in huge amounts of data and make rapid business decisions that possibly stay business to be competitive in real time. Artificial intelligence with business intelligence offers superior and effective outcome in business decision making [4].

Now, the artificial intelligence has extended momentum that is emerging toward ever more complete platforms and solutions that well systematize business intelligence and analytic practices [31], as the growing amount and complication of enterprise-level data have made commercial implementation of artificial intelligence in business intelligence. Moreover, the scholars claimed that using the artificial intelligence and data mining will provide several prospects to improve service offering, productivity, and efficiency [16].

1.2.7 Internet of things (IoT)

The IoT has played a biggest part in today's business operation. For instance, IoT helps e-commerce businesses to keep updated inventory records that reduce the overstock in the warehouses by employing the IoT-enabled sensors [23]. Further, the IoT sensors support to trace the order from the initial placement to final transport that ensures that it should not be fragmented and delivered on time with proper quality. The IoT includes technologies such as RFID, cloud computing, sensors, GIS/GPS visualization, virtual reality, and augmented reality [32]. These technologies are connected to empower the process of business intelligence and decision support system. With the help of IoT, the business intelligence and decision support system focus on target customers for tailored promotion in a way it finds the buying pattern via web surfing to market to their anticipated customers. The IoT devices such as smartphone, tablet, wearable sensors, or industrial sensors can be employed in financial sector for banking activities, in which the business intelligence can be exploited to many banking customers' hidden insights for loans, savings, promotional activities, and transactions.

2 Applications, challenges, and future directions of business intelligence and decision support systems

In general, the business intelligence application includes various components such as database, data warehouse, ETL, data analytical tools, data quarrying, online analytical processing, data imagining, dashboard, customer relationship management, ERP, and other connected components [4]. Scholars proposed augmented business intelligence systems to handle local contextual and smart device–generated data in their surroundings [24]. In recent years, there are various online analytical processing tools which facilitate to explore a dimensional view [33].

When it comes to the challenges, various issues exist in business intelligence and decision support systems. For instance, the scholars highlighted that it is really difficult to identify right workers who possess sufficient capability in both the business issues and analytics [8]. Further, Ishaya and Folarin [11] summarized various challenges in implementing the business intelligent applications across different industries, for instance, legal and ethical issues, issues in collaborating business platforms, inappropriate implementation, and data quality. To surmount the issues in the farming that possesses huge amount of data, the decision support system is proposed so as to take more specific and evidence-based decision by the farmers and stakeholders.

The future direction and avenue in the applications of business intelligence and decision support system will add merits for further developments. Accordingly, scholars suggest that in terms of analytical capability, the traditional applications of business intelligence techniques should be modernized with state-of-the-art technologies for the analytical power [9]. In addition, scholars claimed that the big data analytic solutions necessitate more custom-made software components and superior operational skills than earlier business intelligence applications do [27].

3 Conceptually designed prototypes and architectures for business intelligence and decision support system

This conceptual design proposes the architecture elements based on business intelligence and decision support system that manage, integrate, process, and translate data into valuable information in Figure 3. Accordingly, this includes components such as required skills, various data mining techniques, business intelligence process (data sources, data warehouse, data marts, middleware, and decision support system output), and the conversion process from data to decision in the analytical process. Figure 3 illustrates the conceptually designed business intelligence and decision support system architecture components. The below section detailed each of the components with necessary information.

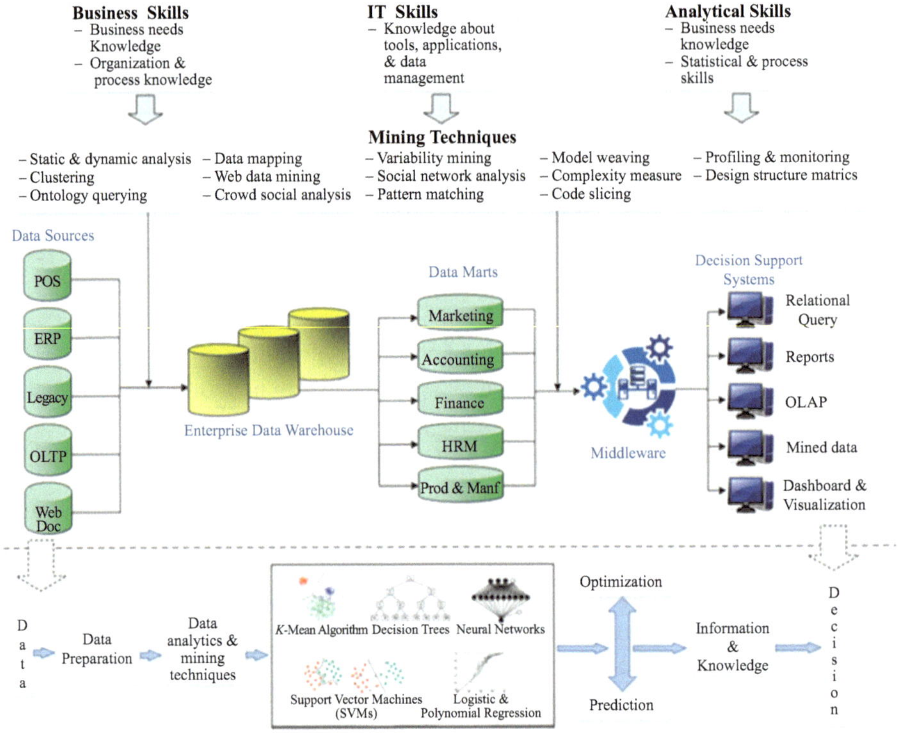

Figure 3: Conceptual design of business intelligence and decision support system architecture.

3.1 Skills requirement

For the better design and development of the business intelligence and decision support system, there are set of skills required for employees. These skills can be from different domains such as business, IT, and analytical power to excel in their position. The business skills are required to understand the business context, organization, and its process, whereas IT skills aim to uplift the capability of the employee's knowledge about the IT tools, modern techniques, applications, various platforms, hardware and software components, data management ability, and so on. IT skills are very much essential to work with today's state-of-the-art technologies, that is, the foundation for today's business intelligence and decision support system. Analytical skills are the abilities of employees related with their cognitive power, expertise, decision-making, and statistical soundness to work as a data scientist or data analyst.

3.2 Data mining techniques

Data mining and analytical techniques are important to process, analyze, and manage big data. There are various types of big data methods and techniques available which analyze large volume of big data. With the recent state-of-the-art development in Internet, technologies, computing, and modeling the big data can be leveraged by analyzing, extracting, mining, visualizing, simulating, and interpreting more informative manner. In addition, the combined power of analytics, machine learning, deep learning, data mining, and text mining certainly helps to explore more insights from big data. These techniques can be used to analyze the data that are stored in the data sources such as Point of Sales (POS), Enterprise Resource Planning (ERP), Online Transaction Processing (OLTP), web documents, and legacy systems. Subsequently, these data can be analyzed and stored into the enterprise-level data warehouse.

4 Business intelligence process

4.1 Data sources

The thousands of millions of business-related data are stored at various data repositories such as POS, ERP, OLTP, legacy systems, and web documents. These data can be structured, semistructured, or unstructured in nature. Afterward, these data can be exacted, minded, and fetched based on the business requirement. For this purpose, various mining and analytic techniques are employed. The processed data in this stage are stored into the enterprise-level data warehouse.

4.2 Enterprise data warehouse

Enterprise data warehouse holds data from various data sources that are processed by data mining techniques. The managing database and data warehousing are referred to as the base of any business intelligence system because these are connected with how data is generated, arranged, kept, dig out, and combined; therefore, the employees can simply access data in a suitable way [5]. The data warehouse is meant that an integrated depository for data has been collected from various heterogeneous databases, while it is used to generate business value with the help of analytical power [33].

4.3 Data marts

The data mart includes sub-databases based on the enterprise's functional areas or department such as marketing, accounting, finance, Human Resource Management (HRM) and production. These data marts are dedicated for specific kind of data for each domain that can be shared from the enterprise data warehouse.

4.4 Decision support system

The decision support systems are able to generate variety of reports, summaries, and output for the stakeholders who are interested. It includes reports, OLAP, mined data, relational queries, dashboard, and visualizations. These summaries and reports will help to find the hidden associations and unseen insights that are essential for managerial decisions. Based on business leaders' and managers' requirement, different types of information can be generated. These can be used for superior business decision-making that in turn stay them in the competitive position.

4.5 Data to decision conversion process

This conversion process aims to convert the data into actionable decisions in the typical business environment. The data passes many stages to become a meaningful information to decisions, in which the data analytics and data mining techniques, namely, k-mean clustering, decision trees, neural networks, machine learning support vector machine, and logistic regressions can be used to forecast the data. These analytic techniques are applied and combined with knowledge to the decision-makers' more timely decision.

5 Conclusion

Today, the enterprise appears to be moving an era governed by big data. The business firms are holding huge amount of data about their customers and routine business operation and managers willing to extract meaningful information from these data. However, there are limitations for the appropriate conceptualization, design, development, and applications of systems in a position to analyze the increasing big data that the enterprises are facing various complexities. Hence, the business intelligence and decision support systems have become an essential element of enterprise's data-driven business decision-making process. This chapter discusses the applications, challenges, design architecture, and technical elements in business intelligence and decision support systems.

References

[1] Arnott, D., Lizama, F. & Song, Y. (2017). Patterns of business intelligence systems use in organizations. Decision Support Systems, 97, 58–68.

[2] Liang, T.-P. & Liu, Y.-H. (2018). Research landscape of business intelligence and big data analytics: a bibliometrics study. Expert Systems with Applications, 111, 2–10.

[3] Alazemi, A. R. & Alazemi, A. R. (2016). Overview of business intelligence through data mining. In: Business Intelligence: Concepts, Methodologies, Tools, and Applications, IGI Global, 49–72.

[4] Aruldoss, M., Lakshmi Travis, M. & Prasanna Venkatesan, V. (2014). A survey on recent research in business intelligence. Journal of Enterprise Information Management, 27(6), 831–866.

[5] Vajirakachorn, T. & Chongwatpol, J. (2017). Application of business intelligence in the tourism industry: A case study of a local food festival in Thailand. Tourism Management Perspectives, 23, 75–86.

[6] Yun, Y., Ma, D. & Yang, M. (2021). Human–computer interaction-based decision support system with applications in data mining. Future Generation Computer Systems, 114, 285–289.

[7] Dunkel, J., Fernández, A., Ortiz, R. & Ossowski, S. (2011). Event-driven architecture for decision support in traffic management systems. Expert Systems with Applications, 38(6), 6530–6539.

[8] Božič, K. & Dimovski, V. (2019). Business intelligence and analytics for value creation: The role of absorptive capacity. International journal of information management, 46, 93–103.

[9] Radenković, M., Lukić, J., Despotović-Zrakić, M., Labus, A. & Bogdanović, Z. (2018). Harnessing business intelligence in smart grids: a case of the electricity market. Computers in Industry, 96, 40–53.

[10] Kao, H.-Y., Yu, M.-C., Masud, M., Wu, W.-H., Chen, L.-J. & Wu, Y.-C. J. (2016). Design and evaluation of hospital-based business intelligence system (Hbis): A foundation for design science research methodology. Computers in Human Behavior, 62, 495–505.

[11] Ishaya, T. & Folarin, M. (2012). A service oriented approach to business intelligence in telecoms industry. Telematics and Informatics, 29(3), 273–285.

[12] Niño, H. A. C., Niño, J. P. C. & Ortega, R. M. (2020). Business intelligence governance framework in a university: Universidad de la Costa case study. International Journal of Information Management, 50, 405–412.

[13] Chen, H., Chiang, R. H. & Storey, V. C. (2012). Business intelligence and analytics: from big data to big impact, MIS Quarterly, 1165–1188.

[14] Sarabi, E. P. & Darestani, S. A. (2020). Developing a decision support system for logistics service provider selection employing fuzzy MULTIMOORA & BWM in mining equipment manufacturing, Applied Soft Computing.

[15] Zhai, Z., Martínez, J. F., Beltran, V. & Martínez, N. L. (2020). Decision support systems for agriculture 4.0: Survey and challenges. Computers and Electronics in Agriculture, 170.

[16] El-Sappagh, S. H. & El-Masri, S. (2014). A distributed clinical decision support system architecture. Journal of King Saud University – Computer and Information Sciences, 26(1), 69–78.

[17] Miranda, J., Rey, P. A. & Robles, J. M. (2012). udpSkeduler: A Web architecture based decision support system for course and classroom scheduling. Decision Support Systems, 52(2), 505–513.

[18] Pérez-Castillo, R., Ruiz, F. & Piattini, M. (2020). A decision-making support system for enterprise architecture modelling. Decision Support Systems, 131.

[19] Vescoukis, V., Doulamis, N. & Karagiorgou, S. (2012). A service oriented architecture for decision support systems in environmental crisis management. Future Generation Computer Systems, 28(3), 593–604.

[20] Serrano-Jiménez, A., Barrios-Padura, Á. & Molina-Huelva, M. (2018). Sustainable building renovation for an ageing population: decision support system through an integral assessment method of architectural interventions. Sustainable Cities and Society, 39, 144–154.

[21] Fertier, A., Barthe-Delanoë, A.-M., Montarnal, A., Truptil, S. & Bénaben, F. (2020). A new emergency decision support system: The automatic interpretation and contextualisation of events to model a crisis situation in real-time. Decision Support Systems, 133.

[22] Leung, K., Choy, K. L., Siu, P. K., Ho, G. T., Lam, H. & Lee, C. K. (2018). A B2c E-commerce intelligent system for re-engineering the E-order fulfilment process. Expert Systems with Applications, 91, 386–401.

[23] Ilmudeen, A. (2020). Big data, artificial intelligence, and the internet of things in cross-border E-commerce, Cross-Border E-Commerce Marketing and Management, 257–272.

[24] Francia, M., Golfarelli, M. & Rizzi, S. (2020). A-Bi+: A framework for augmented business intelligence. Information Systems, 92.

[25] López-Robles, J. R., Otegi-Olaso, J. R., Porto Gómez, I. & Cobo, M. J. (2019). 30 years of intelligence models in management and business: a bibliometric review. International Journal of Information Management, 48, 22–38.

[26] Tu, Y. & Shangguan, J. Z. (2018). Cross-border E-commerce: A new driver of global trade, Emerging Issues in Global Marketing, 93–117.

[27] Debortoli, S., Müller, O. & Vom Brocke, J. (2014). Comparing business intelligence and big data skills. Business & Information Systems Engineering, 6(5), 289–300.

[28] Larson, D. & Chang, V. (2016). A review and future direction of agile, business intelligence, analytics and data science. International Journal of Information Management, 36(5), 700–710.

[29] Gupta, C., Jain, A. & Joshi, N. (2018). Fuzzy logic in natural language processing–a closer view. Procedia computer science, 132, 1375–1384.

[30] Gupta, C., Jain, A. & Joshi, N. (2019). DE-ForABSA: a novel approach to forecast automobiles sales using aspect based sentiment analysis and differential evolution. International Journal of Information Retrieval Research (IJIRR), 9(1), 33–49.

[31] Agrawal, A., Arora, R., Arora, R. & Agrawal, P. (2021). Applications of artificial intelligence and internet of things for detection and future directions to fight against COVID-19. Emerging Technologies for Battling Covid-19, 324, 107.

[32] Li, X. (2018). Development of intelligent logistics in China, Contemporary Logistics in China, 181–204.

[33] Loshin, D. (2013). Data warehouses and the technical business intelligence architecture, Business Intelligence, 105–118.

[34] Ilmudeen, A. Design and development of IoT-based decision support system for dengue analysis and prediction: case study on Sri Lankan context. In: Healthcare Paradigms in the Internet of Things Ecosystem. Elsevier, 363–380.

Andualem Walelign Lale

Business intelligence implementation in different organizational setup evidence from reviewed literatures

Abstract: The work aims to investigate business intelligence (BI) implementation in different organizational setup evidence from empirical literature. The chapter also intended to appeal to both professionals who deal with situations involving change at work on a daily basis, and scholars in the field by contributing toward the knowledge regarding BI technology. This research collects, synthesizes, and analyzes 20 articles on a variety of topics closely related to BI published from 2004 to 2020 in 10 leading information systems (IS) journals. Whereas many studies focus on the role of BI in organizations, few studies state the effectiveness of BI implementation in different organizations' setup through considering empirical findings of different researches. The result reveals that surviving without implementing BI can be very challenging for organizations. Moreover, the overall concept of reviewed literature concluded that effective implementation of BI is subject to strategic mission of the organization and available infrastructure subject to the overall environment.

Keywords: business intelligence; management information system; information technology

1 Introduction

Organizations tend to own an incredible bulk of data. However, much data is deprived in quality or inappropriate whether there has been a large investment in information technology (IT) within an organization. Business intelligence (BI) can, however, help in bringing considerable amounts of information that is useful in a manner that is accurate and timely; such systems, therefore, can enhance decision-making processes [1]. From the emerging results herein, Bestman and Nwanyi [2] appropriately concluded that components of BI strategies have been established as data mining, online analytical processing, querying systems, and reporting systems which significantly relate to quick decision-making as well as time-saving.

Andualem Walelign Lale, Management Department, College of Business and Economics, Jinka University, Ethiopia, e-mail: andualemwalelign26@gmail.com

https://doi.org/10.1515/9783110713633-010

The term "business intelligence" was propagated during the 1990s, well-thought-out as a term that incorporates a wide variety of processes and software used to collect, analyze, and disseminate data, all in the interest of better decision-making [3]. For long time, management information systems (IS) have been supporting organizations in their different operational activities. To respond quickly to changes that take place on the working environment, organizations need management IS that would help effective reaction. Main tasks of BI systems include intelligent exploration, integration, aggregation, and a multidimensional analysis of data originating from various information resources [4]. Thus, BI is an IS-led application that integrates the process and technology to lead to the decision-making for managers and end users [5]. It plays a significant role in analyzing the business environment and providing decision-making in achieving competitive advantages emerging from uncertain often changes within the environment.

Nowadays, data is not only important and valuable to the organization but recognized as necessary to spike the organization's performance and success. As a result, many organizations spend a considerable amount of investment toward obtaining faster accurate information on a real-time basis [6]. A typical architecture for supporting BI within an organization consists of four stages: operational data sources, data integration, data storage, and data presentation. Operational data could be derived from Enterprise resource planning (ERP) applications, Customer relationship management (CMR) applications, manufacturing execution system applications or other legacy systems in organizations, all of which provided data resources for BI [7].

While developing BI system, the role of the organizational culture has been taken into consideration. According to Bach et al. [8], organizational culture settings are more appropriate for achieving higher level of BI maturity. Considering the nature of different organizational culture types, the most probable explanation of the results is that while the investments in BI technology are important, achieving the overall high level of BI maturity go hand in hand and with some organizational culture characteristics which can in turn result in improved organizational performance. Technology improvements and cost reductions have allowed BIS to be extended to the entire set of organizational stakeholders to provide information for various forms of decision-making [9].

Firms should stimulate the development of opportunities that combine inventions generated internally and externally through efficient knowledge transfer and protection of intellectual property, and improved protection against imitation by rivals. BI thus attempts to use the information traffic achieved with firms' current technological integration to achieve interoperability and organization of the flow of information between applications. This means that BI seeks to enable communication not only among the firm's different applications but also with other stakeholders who may form part of the business system [10]. The chapter makes a contribution to development and understanding in the field of implementation of BI and appreciation of the impacts of particular practices upon success for organizations.

Despite the inconsistencies that were identified, the literature review shows that the particular implementation factors result in significant levels of success in the implementation of a BI system. This chapter moreover offers a number of implications for practice, especially for BI stakeholders who are involved in planning, reviewing, or implementing BI.

The compilation above, cited from the literature, offers a basis for considering the range of factors of success and effective implementation of BI. Additional analyses, however, undertaken with the aim of uncovering clear and obvious gaps within the relevant literature have made it apparent that there has been a lack of deep consideration given to the factors that have a bearing on implementation. Likewise, there seems to be little explanation put forward of the particular tactics that may be employed in implementing such systems and to measure whether organizations were effective or not while adopting BI system. Last of all, a further noteworthy observation was that, from the quoted implementation factors, there was a lack of viewpoint taken on user characteristics and work-based learning. Studies by Fetzner et al. [11] emphasize the importance of aligning the organization and the individuals, seeking rational ways to influence and foster the necessary changes in attitudes, behavior, and views or perceptions by adopting actions designed to inform, stimulate awareness, facilitate communication, and take advantage of the influence of key individuals.

As stated by Azeroual and Theel [12], using BI in organizations' operations not only brings advantage but might also have disadvantages. The start-ups have the task, before and during the use of BI, to recognize their benefits and disadvantages and to discuss them openly. The benefits that are offered must be identified and taken so that they lead to both operational and strategic advantages.

Advantages of BI	**Disadvantages of BI**
Functions for targeted and quick research that allow you to find and present relevant information in a current and consistent form	Technology problems (problems of hardware technologies, development systems, and BI software product itself)
Preparation of access to information according to factual and problem-related, possibly multidimensional criteria	Development problems (engineering problems due to difficulties with the task area, with the tools, and with the developers)
Illustration of tasks and problems using meaningful, realistic models	Application problems (complexity of application areas)
Presentation of results in understandable forms of presentation with multimedia techniques (data, texts, graphics, images, and language)	Qualification problems (for all participants). Problems of false expectation (wrong assessment of BI technologies)

Source: Azeroual and Theel [12].

2 Method

The work identified the BI investigations covered in the literature by conducting a systematic literature review. For the review process, the authors use the qualitative approach. Three stages were involved in the process, where the first stage is an identification of articles. The second stage is preparing the analysis and the final stage is finding of analysis and write-up of the result. In order to gather information for analyses, the works of variety of empirical literatures organized by different authors were utilized as an input; moreover, to examine the effectiveness of BI implementation in different organizations, the project. The papers selected in this study and the area where these papers focus are summarized and discussed under the findings.

3 Findings

This chapter aims to investigate BI implementation in different organizational setups. There is a lack of research on the effectiveness of BI implementation. To examine the initiation of organizations regarding BI, 20 researches across the world were assessed. The study by Bader [13] aims to develop a readiness assessment framework based on the critical success factors (CSFs) of BI in Gaza Strip environment. Results showed that top seven factors were top management support, vision and planning, available data quality, resource allocation, appropriate team skills, IT governance, and continuous improvement are the major critical factors essential for BI adoption. Based on the findings, organizations in Gaza are strongly recommended to assess CSFs before starting BI implementation process. Future empirical studies are also recommended to check the relationship between CSFs and BI success. Thus, the study clearly reveals that before implementing BI, the organizations should carefully identify factors that can potentially influence the implementation process.

The work by Figueroa et al. [14] identified BI as an innovative technological way to influence corporate entrepreneurship. This study proposes that BI influences company performance directly and indirectly: directly because it permits innovative development of entrepreneurship dependent on task interdependence in the organization, and indirectly through effective knowledge management, efficient organizational learning processes, and increased technological capabilities in the firm. All of these improvements materialize in data, business processes, and applications, which are in turn innovative forms of entrepreneurship.

In addition to this according to Venuturumilli et al. [15], BI is crucial for retail organizations. While retailers are spending a significant portion of its IT budgets on BI, the actual benefits derived from these tools need to be explored. The study focuses on the organized food and groceries retail, and explores benefits of BI. Data collected from senior marketing executives and managers from six organized food and groceries retail was analyzed using exploratory factor analysis, confirmatory factor analysis, and structural equation modeling. The result shows that BI leads to access to data quality, improved managerial effectiveness, improved operational effectiveness, improved customer orientation, and improved organizational efficiency. The structural equation model also suggests a significant relationship between BI and data quality on organizational performance.

The rising dynamics of business contributed to the need for growth in speed in making business decisions, which tempted the evolution of a new concept in terms of BI that has been made in the form of right-time BI tools. As a new concept that has only recently emerged from the theoretical framework, right-time BI concept blurs the distinction between rational strategic and operational decision-making, allowing the operational level of decision-making to use platforms and tools that were until recently reserved for strategic decision-making. BI greatly improves business decision-making, in particular through the reduction of latency in business decision, which was confirmed by conducted primary research. Its general conclusion is that there is cohesion between performance and the successful application of BI tools [16].

Abai et al.'s [6] study aims to discover the integrated implementation factors of BI and analytics (BI&A) in managing organizational performance, particularly for organizations of the public sector. The study was carried out through interviewing experts, in order to identify the essential factors for BI implementation. The result of the study revealed four integrated factors of the BI implementation, such as skill, documentation, visualization, and work culture. BI solution can provide concrete information regarding the trends and needs of the citizen, allowing for local government organization services optimization and proper knowledge gap discovery if that is the case [17].

Sangar and Iahad [18] also identified that as more organizations move toward intelligent IT infrastructure, nowadays, BI Systems (BIS) become a more widely used IT solution. The study developed a conceptual research framework to identify factors that are critical in BIS implementation. The researchers classified critical factors into two major categories, that is, in the implementation stage and in post-implementation stage. Thus, adequate IT infrastructure, hardware, software, and networks will affect the quality of the BIS and are crucial to its implementation stage and user training and education is an important factor of a BIS success during post implementation stage.

Taking the BI systems specifics into consideration, Olszak and Ziemba [4] present a suggested methodology of the systems creation and implementation in organizations. Hence, in this context, the approach to be used while building and implementing the BI involves two major stages that are of interactive nature, that is, BI creation and BI "consumption." Although basing on standard components shortens time required to build BI, each implementation necessitates adjusting of a particular system to specific requirements of an enterprise. While creating BI for an organization, the three contexts of the technology–organization–environment frameworks were used as the foundational framework of Lautenbach et al. [3]. Organizations wishing to promote fact-based decision-making through greater BI usage are specifically encouraged to consider data-related infrastructure capabilities, top management support, and external market influence, and it is expected that this, in turn, will lead to increased organizational performance.

BI is also important in the operation of small and medium-scale enterprises (SMEs). BI has been recognized as the business imperative for decision-making in conducting change adaptation in SMEs. Because BI is the best conductor of information management and decision making, which generates the business potency. As stated by the study of Ali et al. [19], BI and SMEs are two distinctive research domains but greater interaction between these two entities can offer effective learning from each other. Findings of the literature review suggest enhancing capability of SMEs and new innovation of BI, which may affect each other. In essence, the study has recognized two issues from literature analyzed such as new decision for conducting new changes and organizational entity of SMEs, which raised the issue of new innovation of BI. As a result, BI learns from SMEs. Moreover, start-up companies can be benefited from BI in their decision-making process. Azeroual and Theel [12] assessed the effects of using BI systems on an excellence management and decision-making process by start-up companies. The most frequently mentioned problems with BI use in start-ups according to this case study are the lack of data quality and the lack of data integration from the previous systems. The lack of data integration shows that the start-up companies are not aiming for new isolated solutions with BI, but rather clear, company-wide BI solutions [20, 21]. There is, therefore, a need to catch up in start-ups in the preparation and the matching content requirement profiles for BI solutions.

Nazri et al. [22] identified impacts of BI adoption on organizational performance in the context of higher education institutions (HEIs) in Malaysia in terms of financial gains and customers' management. Based on these arguments, the BI adoption by HEIs in Malaysia is presumed to lead to nonfinancial and financial measures. Finally, the study developed a conceptual framework that indicates the interdependence of BI adoption with organizational performance dimension of financial gain, customer management, internal process, and learning and growth in various frames of technology [23, 24].

BI can also have significant contribution in service delivery system of hospitals. Topaloglou and Barone [25] sought to grasp lessons from a hospital BI implementation organizations. This chapter describes the information needs in a hospital setting and the journey of delivering a BI program. The creation of the hospital BI program was a result of a partnership with other disciplines such as finance, decision support, and clinical operations, and it was a tremendously educational experience to all. The lessons from the BI journey presented in this chapter are drawn from a specific experience in healthcare but in our opinion, they are broadly applicable to organizations that rely on evidence-based decision-making in resource-constrained

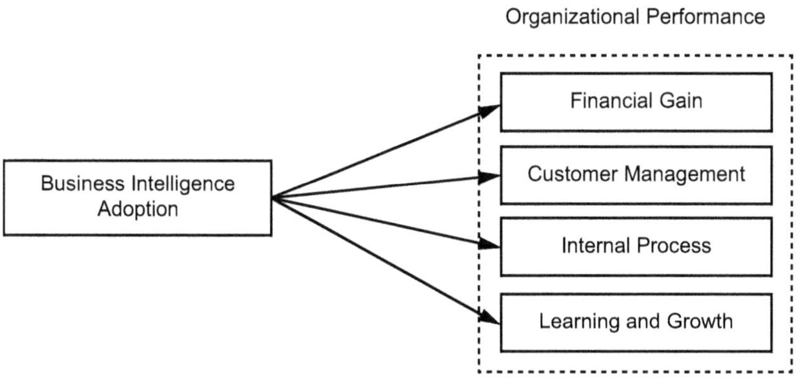

environments. Similarly, Ashrafi et al. [26] aim to bring the reader up-to-date with the current literature on two basic topics such as BI and healthcare delivery and form the basis for the justification of research on the impact of BI on healthcare delivery in the USA. BI tools provide solutions that help healthcare providers effectively manage population health. To survive in a competitive market, healthcare providers need a strong BI foundation to correlate, analyze, and glean insight from financial and operational data. Careful deployment of health analytic tools can allow health systems, hospitals, and individual clinical staff to maximize the value of clinical and administrative data, in many cases without extensive investment in additional IT infrastructure.

By adapting proven information management strategies, healthcare efficiency can be enhanced, waste and redundancy can be abated, patient safety and outcomes can be improved, and public health efforts can be more effectively supported [27]. In the competitive marketplace of today, it is vital for healthcare organizations to adopt strategies and IS which lead to sustainability, improved quality, and competitive position. Information quality is found to be a key differentiator in the ability for learning to influence competitive performance [9].

Finally, the study conducted by Ismail and Al-Assa'ad [28] aims to test the effect of organizational intelligence on organizational agility, which was applied on workers at top and middle management levels. The researcher finds that if the organizations were more interested in strategic vision, "to determine the path adopted by the organization to achieve its mission under the surrounding environmental conditions and competition and the exploitation the highly experienced individuals and their insight into the future directions of the organization and its strategic objectives." Therefore, organizations reach into high levels of sensing agility in the company, that is, increase their ability to keep up with changes in the regulatory environment of customer preferences, the movements of new competitors, and modern technological means by responding well to changes in taking action and generate new insights in the restructuring of organizational behavior.

Generally, through considering the findings of reviewed literature, the major findings of some selected previous works were summarized under the following table:

Title	Major findings and tools to implement BI
Organizational readiness toward business intelligence implementation case study: Ministry of Education & Higher Education – Gaza [13]	Top management support, vision and planning, available data quality, resource allocation, appropriate team skills, IT governance, and continuous improvement are the major critical factors essential for BI implementation
Business intelligence: an innovative technological way to influence corporate entrepreneurship [14]	BI influences company performance directly and indirectly in an innovative way
Evaluating the impact of business intelligence tools on organizational performance in food and groceries retail [15]	Business intelligence leads to access to data quality, improved managerial effectiveness, improved operational effectiveness, improved customer orientation, and improved organizational efficiency.
Integrating business intelligence and analytics in managing public sector performance: an empirical study [6]	Skill, documentation, visualization, and work culture are effective tools to adopt BI in an organization
Critical factors that affect the success of business intelligence systems (BIS) implementation in an organization [18]	Adequate IT infrastructure, hardware, software, and networks, user training, and education is an important factor of a BIS successful implementation
Approach to building and implementing business intelligence systems [4]	Building and implementing the BI involves two major stages that are of interactive nature, that is, BI creation and BI "consumption"
Factors influencing business intelligence and analytics usage extent in South African organizations [3]	While creating business intelligence for an organization, the three contexts of the technology–organization–environment (TOE) frameworks were used as the foundational framework
Analysis of interaction between business intelligence and SMEs: learn from each other [19]	Findings of the literature review suggest enhancing capability of SMEs and new innovation of BI, which may affect each other
The effects of using business intelligence systems on an excellence management and decision-making process by start-up companies: a case study [12]	The most frequently mentioned problems with BI use in start-ups according to this case study are the lack of data quality and the lack of data integration from the previous systems
The impact of business intelligence adoption on organizational performance among higher education institutions in Malaysia [22]	The study developed conceptual framework that indicates the interdependence of business intelligence adoption with organizational performance dimension of financial gain, customer management, internal process, and learning and growth

(continued)

Title	Major findings and tools to implement BI
Lessons from a hospital business intelligence implementation [25]	The creation of the hospital BI program was a result of a partnership with other disciplines such as finance, decision support, and clinical operations, and it was a tremendously educational experience to all
The impact of organizational intelligence on organizational agility: an empirical study in Syrian private banks [28]	Strategic vision is important to implement business intelligence
The impact of business intelligence on healthcare delivery in the USA [26]	To survive in a competitive market, healthcare providers need a strong BI foundation to correlate, analyze, and glean insight from financial and operational data
Business intelligence in college: a teaching case with real life puzzles [29]	Real-life puzzles are a powerful method to teach BI. Real-life problems are interesting for the students, real-life data create a rich source for problem solving using BI, and real-life solutions provide a strong feeling of satisfaction when the answer is obtained. The researcher argues that the puzzle approach works because it imitates the BI process.

3.1 Implications of previous works for business intelligence practice development

The rapid increase in data volumes in companies has meant that momentous and comprehensive information gathering is barely possible by manual means. BI solutions can help here. They provide tools with appropriate technologies to assist with the collection, integration, storage, editing, and analysis of existing data. While almost only large companies were interested in this topic a few years ago, it has meanwhile also become necessary for start-up companies, and so the market for BI has been growing for years. Well-functioning BI project coordination requires interdisciplinary competence between specialist knowledge (above all BI know-how) and leadership in the form of assertiveness, communication, and coordination skills.

According to scholar's previous works, an IS can be defined as a "set of interrelated components that collect (or retrieve), process, store, and distribute information to support decision-making and control in an organization." BI is an important area of IS due to the potential to analyze and solve problems of the company [30]. For having successful BI practice in an organization, points stated in Figure 1 is critical [17].

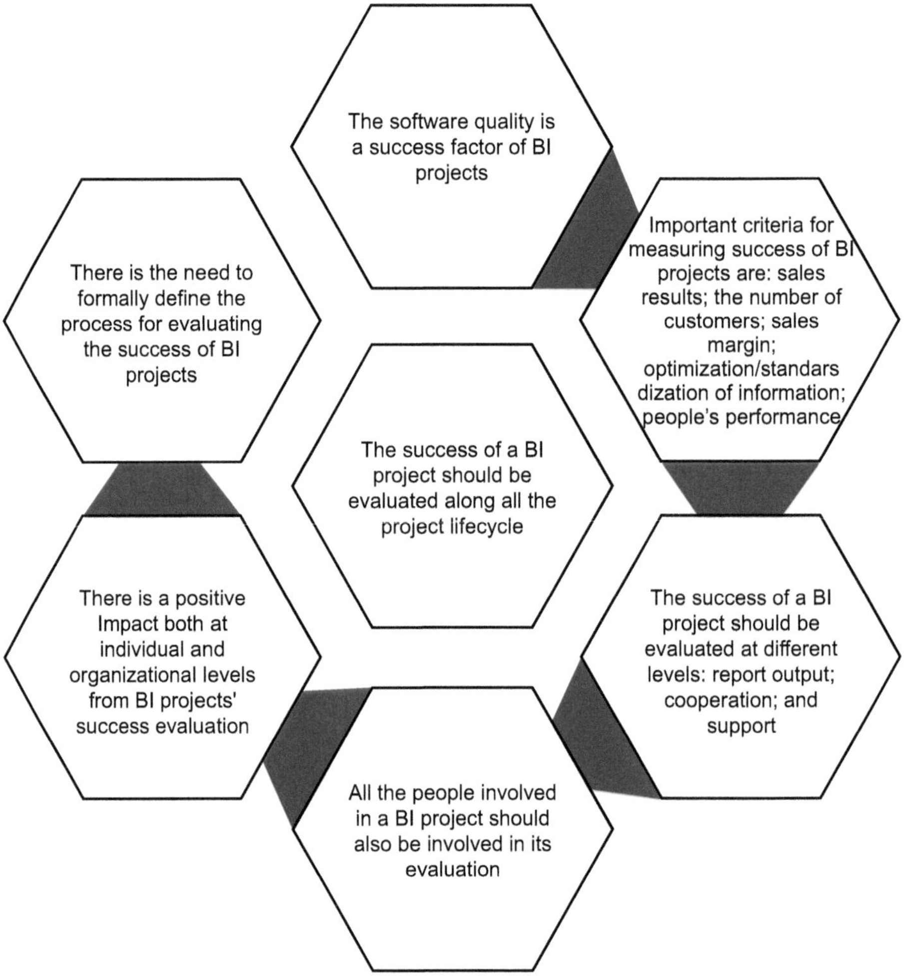

Figure 1: Basic elements for effective business intelligence practice.

4 Challenges in business intelligence implementation

The implementations of BI&A within organizations are viewed as difficult and challenging because they extend beyond simple software and hardware implementations and are more complicated to deploy and run. BI&A implementation projects are complex and often involve lengthy integration processes [3].

Many areas such as education, hospitality, small and medium enterprises (SME) and healthcare have seen the potential of BI to accelerate the business process as to meet the objectives of organizations. The purpose of BI also has given the benefit not only to the organizations or to industries but also to the customers as an external entity that supports the growth of business, for example, in the hospitality industry. Different papers discover problems that become challenges for the implementer to be successful in BI executions. The review research by Hasan et al. [31] found that factors that pose challenge to the organizations can be categorized under organization dimension (limitation of funding (high cost in setting the BI infrastructure and hiring BI experts), process dimension (lack of expertise, training, and user acceptance), and finally the technology dimension (data management issue).

5 Conclusion

Modern BI is about the process of turning data into actionable information, using an assortment of tools, techniques, and applications. Although BI, or its predecessor decision support systems) has been applied in the industry for about half a century, it has only recently been taught in business schools. This research has involved the review of literature published from 2004 to 2020 and discovered that effective practice of BI is a success factor for every organizational setup. BI solution can provide concrete information regarding the trends and needs of the citizen, allowing for organization services optimization and proper knowledge gap discovery for challenging competitive edge. Organizations today collect enormous amounts of data from numerous sources [32]. The use of BI to collect, organize, and analyze this data can add great value to a business. There are two perspectives of BI: technological and organizational. Technological means a system that takes data and transforms it into various information products, while organizational means an umbrella term for decision support. The overall concept of reviewed literature concluded that effective implementation of BI is subject to strategic mission of the organization and available infrastructure subject to the overall environment. Thus, for future investigation and successful BI implementation, the author depicted the major finding of the reviewed article by the following diagram:

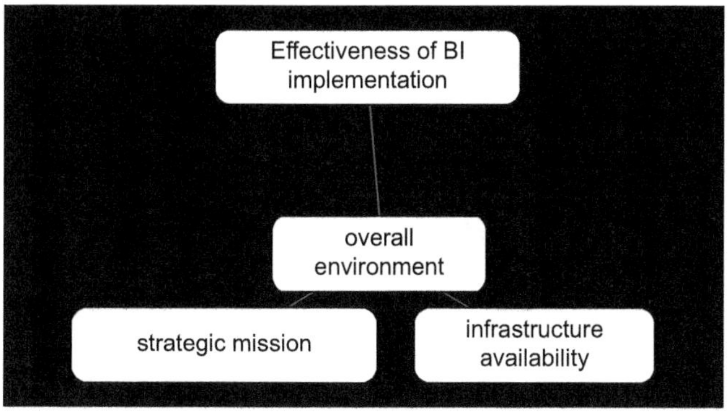

6 Research limitations and recommendation

This research chapter is not without limitations. First, the study can be considered as only looking at the level of the implementation of BI in organization setup; as such, lots of other themes of research related to the implementation of BI systems are overlooked. Second, there has not been exploration of the research paradigms in this study in methodological and theoretical terms. Third, the author noted that several methodologies were either underrepresented or absent from the pool of BI research. Further empirical research of that area could discover other facts in relation to factors and their impact upon success. There ought to be careful consideration and assimilation of these concerns in any further related studies through considering the real-world practice of organizations. With regard to the validity of the emerging theory, it is important to address generalizability, which is the validity of a theory in a setting different from the one where it was empirically tested and confirmed.

There is currently a trend toward the integration of BI systems with existing IS in order to improve decision-making capabilities in organizations. Even though much attention has been paid to the factors influencing the adoption of BI systems, in practice there is still limited research investigating the business value of BI systems in a post-adoption environment. Research recommendations are as useful for researchers as they are for organizations wishing to implement systems of BI successfully. The review of literature, in relation to factors that have an association with research of the implementation of BI, ought to have analysis of the factors that are used most commonly with respect to BI system implementation; this would provide researchers with a path toward proper analysis of what factors lie behind success. This review of literature can also serve as a guide for organizations seeking to take preventative measures for avoiding some of the challenges that are potentially faced while trying to implement a BI system successfully.

References

[1] El-Adaileh, N. A. & Foster, S. (2019). Successful business intelligence implementation: a systematic literature review. Journal of Work-Applied Management, 11(2), 121–132. doi: https://doi.org/10.1108/JWAM-09-2019-0027.

[2] Bestman & Nwanyi. (2019). Business intelligence system strategies and organizational success in public hospitals in rivers state, Nigeria. European Journal of Business and Innovation Research, 7(2), 1–21.

[3] Lautenbach, P., Johnston, K. & Adeniran-Ogundipe, T. (2017). Factors influencing business intelligence and analytics usage extent in South African organizations. South African Journal of Business Management, 48(3), 23.

[4] Olszak, C. M. & Ziemba, E. (2007). Approach to building and implementing business intelligence systems. Interdisciplinary Journal of Information, Knowledge, and Management, Volume 2.

[5] Miah, S. J. (2014). A demand-driven cloud-based business intelligence for healthcare decision making. Handbook of research on demand-driven web services. Theory, Technologies, and Applications: Theory, Technologies, and Applications, 324.

[6] Abai, N., Yahaya, J., Deraman, A., Hamdan, A., Mansor, Z. & Jusoh, Y., (2019). Integrating Business Intelligence and Analytics in Managing Public Sector Performance: An Empirical Study International Journal on Advanced Science, Engineering and Information Technology.

[7] Hou, C. K. (2015). Using the balanced scorecard in assessing the impact of BI system usage on organizational performance: An empirical study of Taiwan's semiconductor industry. Information Development.

[8] Bach, M., Jaklic, J. & Vugec, D. (2018). Understanding impact of business intelligence to organizational performance using cluster analysis: does culture matter?. International Journal of Information Systems and Project Management, 6(3), 63–86.

[9] Woodside, J. M. (2010). Business intelligence and learning, drivers of quality and competitive performance. ETD Archive, 151.

[10] Richards, G., Yeoh, W., Chong, A. Y.-L. & Popovic, A. (2014). An empirical study of business intelligence impact on corporate performance management. PACIS Proceedings. Paper 341. http://aisel.aisnet.org/pacis2014/341.

[11] Fetzner, M., Amélia, D. M. & Freitas, H. (2011). Business Intelligence (BI) implementation from the perspective of individual change. JISTEM – Journal of Information Systems and Technology Management, 8(1), 25–50.

[12] Azeroual, O. & Theel, H. (2018). The effects of using business intelligence systems on an excellence management and decision-making process by start-up companies: A case study. International Journal of Management Science and Business Administration, Volume 4(Issue 3), March 2018, Pages, 30–40.

[13] Bader, A. (2017). Organizational Readiness Toward Business Intelligence Implementation Case Study: Ministry of Education & Higher Education – Gaza.

[14] Figueroa, R. & Martin-Rojas., R., and Garcia-Morales, (2017). Business Intelligence: An Innovative Technological Way to Influence Corporate Entrepreneurship, DOI: 10.5772/Intech open, 70538.

[15] Venuturumilli, S., Peyyala, P. & Rao, P. (2016). Evaluating the impact of business intelligence tools on organizational performance in food and groceries retail. Journal of Economics and Business Research /Articles, 22(2).

[16] Dukic, B., Bara, D. & Dukic, S. (2016). Impact of Right-time business intelligence tools on efficiency in decision-making. Technical Journal.

[17] Teixeira, R., Afonso, F., Oliveira, B., Portela, F. & Santos, M. (2019). Business Intelligence to improve the quality of Local Government Services, Case-Study in a Local Government Town Hall.

[18] Sangar, A. &. & Iahad, N. (2013). Critical factors that affect the success of Business Intelligence Systems (BIS) implementation in an organization. International Journal of Scientific & Technology Research, 2(Issue 2).

[19] Ali, S., Miah, S. J. & Khan, S. (2017). Analysis of interaction between business intelligence and SMEs: Learn from each other. Journal of Information Systems and Technology Management, 14(2), May/Aug., 2017, 151–168.

[20] Ahuja, R., Vats, K., Pahuja, C., Ahuja, T. & Gupta, C. (2020). Pragmatic analysis of classification techniques based on hyper-parameter tuning for sentiment analysis International Semantic Intelligence Conference (ISIC'21), Delhi, pp. 453–459, 2021.

[21] Gupta, C., Chawla, G., Rawlley, K., Bisht, K. & Sharma, M. Senti_ALSTM: Sentiment analysis of movie reviews using attention-based-LSTM. In Proceedings of 3rd International Conference on Computing Informatics and Networks: ICCIN 2020 (p. 211). Springer Nature.
[22] Nazri, S., Ashaari, M., Iskandar, Y. & Bakri, H. (2019). The impact of business intelligence adoption on organizational performance among higher education institutions in Malaysia, Advances in Economics, Business and Management Research, Series volume number 141.
[23] Agrawal, P., Madaan, V., Roy, A., Kumari, R. & Deore, H. (2021). FOCOMO: Forecasting and monitoring the worldwide spread of COVID-19 using machine learning methods. Journal of Interdisciplinary Mathematics, 24(2), 443–466.
[24] Gupta, C., Jain, A. & Joshi, N. (2019). DE-ForABSA: a novel approach to forecast automobiles sales using aspect based sentiment analysis and differential evolution. International Journal of Information Retrieval Research (IJIRR), 9(1), 33–49.
[25] Topaloglou, T. & Barone, D. (2015). Lessons from a Hospital Business Intelligence Implementation, Proceeding from CAiSE.
[26] Ashrafi, N., Kelleher, L. & Kuilboer, J.-P. (2014). The impact of business intelligence on healthcare delivery in the USA. Interdisciplinary Journal of Information, Knowledge, and Management, 9, 117–130. Retrieved from http://www.ijikm.org/Volume9/IJIKMv9p117-130Ashrafi0761.pdf.
[27] Ferranti, J., Langman, M., Tanaka, D., McCall, J. & Ahmad, A. (2010). Bridging the gap: leveraging business intelligence tools in support of patient safety and financial effectiveness. Journal of American Medical Informatics Association, 17(2).
[28] Ismail, H. & Al-Assa'ad, N. (2020). The Impact of organizational intelligence on organizational agility: An empirical study in Syrian private banks. International Journal of Academic Research in Business and Social Sciences, 10(2), 465–483.
[29] Presthus, W. & Bygstad, B. (2012). Business intelligence in college: A teaching case with real life puzzles. Journal of Information Technology Education, 11.
[30] Fourati-Jamoussi, F. & Niamba, C. N. (2016). An evaluation of business intelligence tools: a cluster analysis of users' perceptions. Journal of Intelligence Studies in Business, 6(1), 37–47.
[31] Hasan, N., Rahman, A. & Lahad, N. (2016). Issues and challenges in business intelligence case studies. Jurnal Teknologi (Sciences & Engineering), 8–2, 171–178.
[32] Mohammad, H. A. (2012). The Impact of Business Intelligence and Decision Support on the Quality of Decision Making, An Empirical Study on Five Stars Hotels in Amman Capital.

Hazik Mohamed

Conceptualization of a modern digital-driven health-care management information system (HMIS)

Abstract: From our understanding of the limitations of the current health management information systems (HMIS) and their various configurations, the modern HMIS system needs to be adaptable, transparent, and interoperability with other independent systems like insurance systems and private health-care records.

This research focuses on the modern HMIS ability to improve the operational efficiency of health-care facilities, generation, and use of medical data for the benefit of its citizens and stakeholders, electronic record-keeping, and the ability to share the data across various systems. Such a system will involve blockchain transactions and the immutable tracking of services extended. In the proposed configuration, the modern HMIS is built to allow data consolidation and improve health-care services and clinical research. The use of data analytics will reduce errors and improve clinical decision making and universal access to critical medical information in real time. Flexibility and the ability to interface and integrate to other systems will play a crucial role in data access and ultimately the cornerstone of an efficient medical record-keeping system. Such a system not only manages the clinical information but it will also manage pharmacy supplies, efficiently dispatch hospital services such as nurse management of the patient, bed transfers, radiography, consultation tracking which will all be integrated into the HMIS modules with real-time visibility via a shared dashboard.

Keywords: big data, electronic medical records, financial disputes, health-care services, interoperability

1 Introduction

With the global demand for health care surmounting the ability of patients to pay for rising costs, the health-care industry appears to be headed for a crisis globally. The number of elderly people versus the number of employable people is increasing as the world population ages. As such although there is an obvious need for elderly health care, there are less salaried working populace to provide financial support for medical costs and health-care insurance. Furthermore, “the cost of providing medical

Hazik Mohamed, Putra Business School/Stellar Consulting Group, e-mail: hazik@putrabs.edu.my/ hazik@stellarcg.com

https://doi.org/10.1515/9783110713633-011

care continues to outpace inflation, with the average global rise in medical care projected to be 4.9% in 2020 and 4.0% in 2019" ([1], p. 12), up from 4.0% in 2018.

As people live longer, the percentage of working people to pay for increasing medical costs is comparatively lower, which is creating a weighty problem for governments. The health-care industry and its businesses have to find innovative ways to make the health care more efficient and less costly for everyone. Governments can create improve supporting subsidies for common aging ailments. Most of all, technology can offer solutions to medical care and become a modern means to accomplish these goals and resolve traditional issues.

One such application is the electronic medical records (EMRs). Most patients move between medical care providers, and in doing so, their medical records are localized by diagnosis of that provider. Each medical record becomes scattered across different organizations if the patient keeps moving, as they do not personally have access to their past records. The framework has always been that medical service providers are the primary stewards of medical records, and are assumed responsible for their record-keeping, and storing the information. From the patient's perspective, much like in consumer finance, the individual uses several service providers and creates multiple bank accounts, possesses several credit cards, obtains a few loans, and acquires a range of assets but no single way to access and control them. The current medical record-keeping systems were not designed to be so far-sighted enough to manage lifetime medical records and interoperable between centralized institutions.

The idea is to combine a social need with a technological enabler: a system that returns control of medical records to the patient who owns them, while empowering them to determine who can have access to their medical history. We discuss and attempt to conceptualize a solution that can distribute access and validate authenticity and permissions via a blockchain system to replace centralized intermediary institutions. The shift for the health-care industry toward such innovation involves personalization of data, which prompt patients to engage in the details of their health care without adding additional burdens on them. The proposed solution simply restores ownership and agency over their own medical records. Blockchain used in health-care systems will provide access control, permissioned auditability, and interoperability between large data management systems. It is expected to revolutionize and drive the digitalization of health care, particularly in medical record-keeping and the sharing of sensitive medical data securely and quickly via a trustworthy, immutable channel.

In this chapter, we discuss the concept of the modern health-care management information system (HMIS) and the ways where technologies like the blockchain can primarily secure personal and confidential patient data giving permissioned access approved by the patient herself/himself to related parties like health-care providers, pharmacies, and health/medical insurers. Patients do not transfer the data to the parties requesting it, but he/she is merely giving them access to view it to enable their work. Thus, this paradigm shift enables the emergence of data privacy and control, releasing sensitive information only to relevant parties like medical practitioners, insurers, pharmacies and researchers while engaging patients with their service providers.

In addition, this concept will be fully integrated with a modern HMIS that is able to improve the operational efficiency of health-care facilities, creation and use of medical data for the benefit of its patients and other stakeholders, and the ability to share the data across various systems securely and efficiently. Such a system will involve blockchain transactions and the immutable tracking of services extended.

For future medical records to be more effective, an artificial intelligence (AI)-driven patient-centric system will be suitable to complement the blockchain EMR-keeping system. In the proposed configuration, the modern HMIS is built to allow data consolidation and improve health-care services and clinical research. The expanding use of data analytics intelligently will reduce errors and improve clinical decision making and universal access to critical medical information in real time.

2 Main pillars of the modern HMIS

The modern HMIS will provide hospitals with a system to benchmark themselves using the EMR adoption model also known as EMRAM's eight stages to achieve meaningful use of health IT and create a paperless patient record environment. Physicians will be able to reach proper diagnoses by asking relevant questions and getting answers sought in order to perform appropriate medical tests to determine causes and remedies for them. It will also prevent "errors of omission" in the front end that may prove critical when trying to prevent "errors of commission" on the back end, that is, during treatment and procedures.

As such, we determine the three main pillars of the modern HMIS in further sections.

2.1 Ease of use

The HMIS will make it easy to enter, manage, and retrieve pertinent patient data whenever required. It will also be able to record prescribed medication by the attending physician, subsequently print prescriptions for dispensation by the pharmacy, issue appropriate medical certificates as well as create any other special reports, as needed.

2.2 Fully integrated

It will be a fully functional EMR and practice management system integrated together. The modern HMIS can schedule medical and specialist appointments, manage unscheduled changes or amendments, and generate performance reports like duration per appointment and clinic revenues with the built-in billing system according to access levels.

2.3 Mobile and accessible

HMIS users can sign up and log in to the system using any device: desktop, laptop, and even tablets. The system will be hardware agnostic and allow for mobility and versatility. HMIS can be accessed as long as there is Internet or Wi-Fi connectivity. If there is Internet, it can be accessed, edit clinical records with predetermined user access based on the employees' job scope. No longer are employees bound to desks to access desktop computers.

3 Structure and features of the modern HMIS

The data structure of the system manages and handles all the data from the different modules, making it a complete integrated system as shown in figure 1. Not only will it manage clinical information but will also manage other administrative functions through an order management module that allows for nurses to manage and prioritize patients and room availability in terms of medical severity. The system will also be integrated to the appropriate billing elements and to other service providers (like the MRI, pharmacy, and radiology) within the medical facility as well. This system will

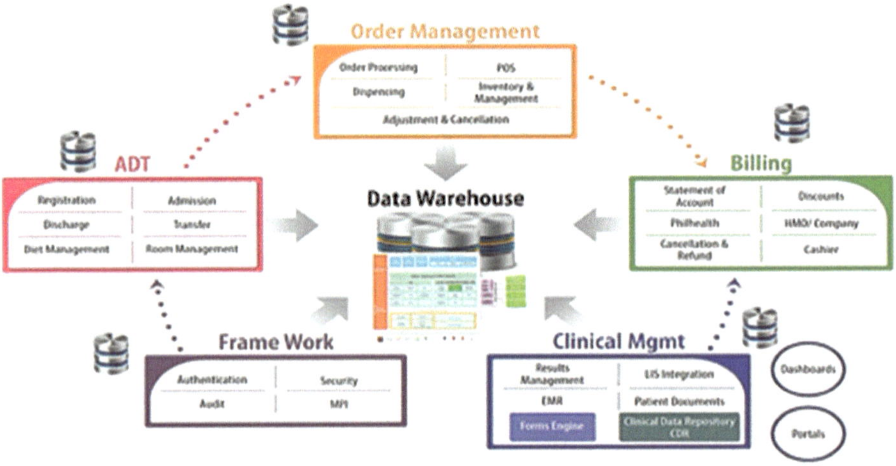

Figure 1: The key structural features to the modern HMIS.

be able to track all the costs for the services and treatments being rendered to each patient to effectively track any procedures, treatments, and/or drug dispensation for the patients and minimize disputes over erroneous billing.

Its architecture includes the various services that will manage its data integrity, security, control, and access. The architecture allows for interoperability for integration to any system that supports third-party connectivity. It will be able to integrate with any nonclinical integration like Enterprise Resource Planning (ERP) finance system, which will allow for full integration, removing the needs to any human intervention. This is important especially for any finance audit, as it will remove any concerns about data manipulations or adjustments.

The main features of the system can be summarized as follows:

1. Generate a full patient record for quick review for consultation, diagnostics, and procedures.
2. Utilize various templates for simple operating procedures or combine them for customized ones including documentation for different specialty practices.
3. Take, upload, and annotate images to patient's medical record at consultation or after diagnosis.
4. Instruct and control prescriptions according to dosages and limitations with the HMIS built-in drugs and medication database.
5. Document and view charts and other visual representations of vital signs and other medical test results. Evaluate and benchmark against trends of laboratory values from a registered database for more informed analyses.
6. Record patient payments and payment type (i.e., cash or credit), monitor revenues, and track outstanding balances with the HMIS built-in billing module.
7. Monitor inventory levels of drugs, treatment supplies, and other items including specialized expandable procedures and services provided by each clinic. Alerts for inventories running low and need replenishment.
8. Create other census and insightful custom reports like the top 10 reasons for consultation or the 10 most prescribed medications in a given period for a strategic analysis.
9. Integrate old medical records of patients with the modern EMR system through enhanced functionalities to ensure full record transfer for a more complete record.

Flexibility is a critical advantage for any new system. The ability to interface and integrate with other systems – existing or new – this interoperability of integrated systems will govern the sharing of critical information via permissioned data access which will ultimately establish the foundation of an efficient medical record-keeping system. Such a flexible and interoperable system not only manages the clinical information but will also assist in managing pharmacy supplies, efficiently dispatch hospital services such as administration management of the patient, nursing needs, bed transfers, radiography, and consultation tracking which will all be integrated into the HMIS modules with real-time visibility via a shared dashboard. The modern HMIS is a system that can track all costs for the services and supplies being rendered to each patient in a transparent manner, essentially eliminating disputes over medical bills and other miscellaneous charges.

4 Benefits of digitalized processes for medical record-keeping

"Electronic records are superior to paper records because they decrease error due to handwriting problems and ease physical storage requirements. Additionally, electronic records simultaneously leverage other error-reducing technologies and render them coherent"[1] [2]. Gunther and Terry believe that EMR models "present significant additional advantages because of their potential to deliver a longitudinal record that tracks all medical interactions by a patient and provide comprehensive data across populations."

On top of data consolidation, longevity, and overall reduction in errors, EMR adoption will be driven by business concerns and structural changes in modern health-care delivery. These four main drivers are a result of:

1. The accelerated shift from inpatient to ambulatory care (and other irregular events) from the need for accurate and efficient flow of patient medical and billing information between different medical care providers, both geographically and organizationally.
2. The increasing need for data analytics for insights and transparency in operational care demands for performance measures, and system administrators' desire for a more sophisticated review and risk management tool.
3. The growth of "shared care" where the patient shares the responsibility of care with the health-care provider which are fragmented by having multiple providers throughout the course of a lifetime. As such, patients must have full access to their own medical records to inform subsequent care providers on other occasions of treatment and/or procedures.
4. Both patients and regulators asking for instances where medical errors or near-fatal mistakes have occurred. Such data are often covered up and difficult to capture without access or voluntary reporting by the providers themselves. Even when regulated to do so, it will be nearly impossible to analyze such data without advanced database systems.

Storing medical records on the blockchain will make private health-care data meddle-free, protected, and scalable across future systems. "The distributed nature of the

1 https://www.ncbi.nlm.nih.gov/pmc/articles/PMC1550638/

blockchain can ease the sharing of data among authorized parties and bridge traditional data silos, dramatically increasing efficiencies and improve coordination of care"[2] [3]. Intuitively, with better coordination between various parties involved (e.g., public, hospitals, insurance companies, and pharmacies), the costs of medical care can be significantly reduced. "Data auditing is improved through the immutable records maintained by the blockchain. The costs associated with blockchain mining can even be offset by offering anonymized metadata rewards for medical researchers"[3] [4].

As in all blockchain-based projects, its full scalability can only be achieved through consensus on protocols for the coding of medical information and interoperability of data exchange across systems. Its success will also "depend upon its ability to keep medical records private while at the same time widely available to medical providers and insurance companies" [5]. Such is the requirement for a digital medical record blockchain initiative to effectively scale from a national blockchain-based EMR to the global level.

5 Utilizing technology for medical record-keeping

Since health care is one of the most regulated industries, deploying blockchain applications can improve trust and accountability between various parties in the health-care domain. "Blockchain technology has led to tremendous solutions for traditional health care domain issues" [6], such as "providing a secure infrastructure and integrated private health records" [7]. Blockchain can be used to "provide secure communication among stakeholders and deliver clinical reports efficiently" [8]. The use of AI and blockchain in different applications for medical operations and health care is one of the most interesting opportunities in current practical research.

Blockchain allows sharing an EMR in a "secure manner since blockchain technology can be extended as a standard for stakeholders" [9]. Digitizing medical records as EMR extends numerous advantages, one of which is protecting the patient's privacy [10] and improving quality of medical care [11]. Striving toward patient-centric services must connect independent systems using a shared decentralized platform (see Figure 3). EMRs contain sensitive and private data and should be kept in a secure and confidential system with controlled access. As such, it has become a "prime target for malicious attacks, such as Denial of Service (DoS), Storage Attacks" [12], and so on. Understanding private and public blockchain use cases will provide ideas for different access control mechanisms to defend against such failures and attacks, to establish a more secure and robust platform for health-care services.

2 https://www.finextra.com/blogposting/13801/blockchain-in-healthcare-make-the-industry- better
3 https://www.pubpub.org/pub/medrec

6 Record management system

An effective EMR management (EMRM) system should address "several major issues related to health care, namely data quantity and quality, system interoperability, fragmented medical data, patient agency, and slow access to medical data" [13]. A set of APIs has to be built for patient database integration in order to provide interoperability. In a blockchain setup, automated data retrieval instructions and viewing permissions can be deployed using smart contracts between patients and providers. Contributing entities control their records by accepting or rejecting new information as illustrated in Figure 2, where node A is a public hospital, node B is a private hospital, node C is a pharmacy, and node D is an overseas medical care specialist.

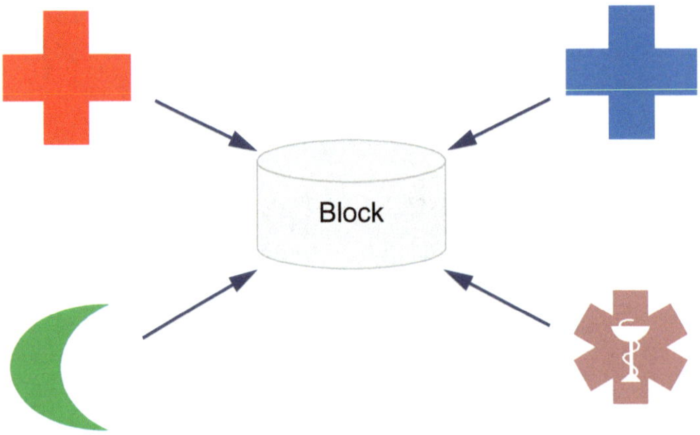

Figure 2: Defining appropriate nodes for validation of new information.

Identity confirmation is done through a DNS-like implementation by linking a specific Ethereum address to a specific patient's ID. An algorithm synchronizes off-chain data exchange between the provider and patient's database, while an authentication module is used to confirm permissions to database access. The proposed system allows data sharing, confidentiality, authentication, and accountability for sensitive information, and provides patients with easy access, an immutable medical history log, and reliable medical information across treatment locations, and health-care providers. A trustworthy EMRM system will be dependent on multiple nodes (entities) for verification of new information and avoids single points of failure. In addition, it must include contract encryption and preserving auditability while improving complexity faced in the existing systems.

6.1 Privacy preserving systems and data exchange

A permissioned-based framework can offer privacy through efficient access, interoperability, and security of all medical records for parties seeking access, be they health-care providers, insurers, or other patient care specialists. A proposed framework should be able to preserve patients' privacy using cryptographic techniques used in blockchain operations whether they are databases or automatable smart contracts to execute the necessary steps or procedures. These permission-based processes allow patients to manage control the privacy of their own medical records whose integrity can be confirmed using query links and cryptographic hashes. "Multi-signature contracts are utilised to control and administrate activity of a specific account, by establishing powers and rights to that specific account. Multi-signature contracts allow for using multiple keys to edit the [medical] data on the chain; thus, the entities that hold all the keys can edit the information, while others can only read or initiate transactions from the ledger" ([14], p. 141). For the immense amount of medical data, the data are stored in a decentralized repository called a "data lake" instead of the blocks on the chain itself. A data lake can store various types of data and is highly scalable due to interoperability. Only verified data are amassed onto the data lake, and as illustrated in Figure 3, only the patient holds the right to give access to his/her medical records data. This encrypted sharing approach allows patients to manage access and to have full control over their own medical records, and it enables multi-institutional access across various types of devices and is particularly helpful for quick data access (especially in historical assessment, etc.) during emergency or urgent treatments.

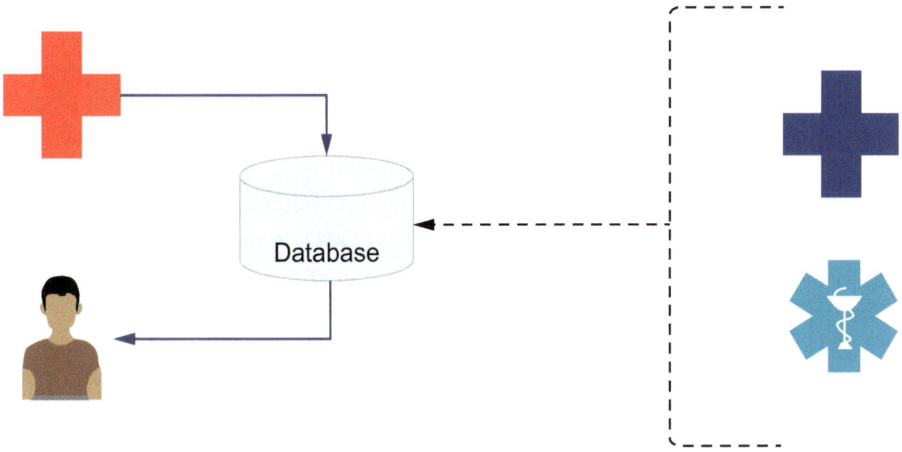

Figure 3: A permissioned-based access to digitized medical records.

Another method to protect data privacy is through utilization of loosely coupled blockchains to enhance the privacy by combining "on-chain verification" and "off-chain storage." Some models use "the distributed databases of hospitals to ensure off-chain storage, while the hash value of medical records is used in a transaction for on-chain verification" ([15], p. 53). Other architectures use "context blockchains, off-chain storage and exchange reference pointers on-chain, token-based permission model and Model-View-Controller (MVC) pattern to provide a comprehensive solution" ([15], p. 54). They use public key cryptography to create and manage health identities.

EMR validity can be guaranteed via an attribute-based signature scheme for multiple authorities on the blockchain for EMR. "In this approach, patients will endorse messages based on specific attributes without disclosing other information. The only parameter needed to access the messages is providing evidence to the verifier" ([16], 11,682). Multiple authorities can distribute patient's public and private keys in favor of a single authority to avoid escrow complications. Insurance records, drug consumption, and treatment records along with other patient information are recorded on a single block upon treatment completion. If the patient goes to another health-care service provider in the future, only authentication of the patient's identity is required for the patient to grant access to historical data to begin the new treatment.

7 Mobile application architecture

For mobility of access, the mobile application can be designed to collect data from patients in a secure manner, after which it will synchronize with cloud services to share the information with health-care providers and health insurance companies. Mobile-based (or DApps) should provide health data and medical records sharing solutions through user-centric models but cybersecurity remains a major concern, so cloud services that are used need to be carefully examined. Some prototypes and designed mobile applications have used "membership service (supported by blockchain) and a channel formation scheme to enhance identity management service and privacy protection by utilizing permissioned blockchain characteristics" [17]. In such DApps, the mobile architectures utilize IoT (Internet of things) devices such as sensory medical and wearable devices to collect patient information, on top of manual data input. The large amount of medical data accumulated via the mobile platforms or synced wearables are handled by batching and tree-based data processing methods.

For illustrative purposes, let us consider this typical scenario. Patients typically only decide to see their doctor, when pain or symptoms become unbearable, or an accident has occurred. When this happens, they have to schedule an appointment or rush to the emergency room, depending on the severity. During consultation or registration, they must remember the pertinent details of their medical history, such as drug allergies, past operations, and existing medical conditions. In such situations, patients may forget or too anxious to remember critical information crucial to a medical assessment. Instead with a blockchain data lake, data can be shared via patient permission. In addition to interoperability of information and systems, patients can be sensor-equipped to monitor their movements (or nonmovement) within their homes or domiciles via a DApp enhanced by wearables and IoT devices to determine when something has gone wrong, early stage on, especially in the case of elderly persons.

For other scenarios concerning scheduled visits, doctors can record patient visits and have external apps retrieve demographic information from the shared patient's medical record (EMR) when the patient arrives. Future appointments can be booked via these DApps and notifications of visits recorded in the clinicians' schedule, typically maintained in the EMRM. When exchanging more clinically related data, interoperability between DApps with EMRMs will have to resolve semantic interoperability issues through standard coding terminologies to alert potential drug allergy reactions for a prescription, for instance.

A diagnosis recorded during other medical encounters can be codified and sent back into an EMRM system, which is shared onto a cloud-based personal EMR. Similarly, lab and radiology reports can be codified for personalized EMRs and shared whenever required across systems where permission is enabled only by the patient.

8 Modern HMIS in the context of national health-care policy

Modern HMIS "present a valuable opportunity to improve health surveillance and evaluate service provision potentially leading to improvements in the management and the promotion of public health" ([18], p. 1). Public health research includes preparedness for pandemics (e.g., bird flu, COVID-19, and swine flu outbreaks), cancer, maternal health, infectious diseases such as leprosy, and sexual health. It also "incorporates evaluations of systems already in place in urban and/or rural regions ranging from primary to tertiary care as well as across different health care providers, such as nongovernmental organizations" ([18], p. 4).

Table 1: Tasks and benefits of national EMR systems to public health.

National HMIS policy	
Disease surveillance and outbreak monitoring	**Health-care system continuous improvement**
– Monitor trends – Planning for outbreaks – Identify transmissions and action plans – Action plan accountability (closed-loop governance)	– Sharing/transfer of data between hospitals/insurance/government ministry systems – Warning triggers for patient safety and alerts to improve clinical decision making – Updated records for interoperability with DApps and wearables

A key consideration of public health in national health-care policy is the utilization of EMR for disease surveillance and outbreak monitoring systems. EMRs coupled with contact-tracing devices can help to identify sudden outbreaks and control high-risk areas to manage pandemics. Movement control in such areas can prevent further spread of infections and assist in the coordination of containment measures, which are vital in the current public health climate due to the ease of mobility. Another key utilization of modern HMIS (as shown in Table 1) is the pursuit of continual improvement for public health-care systems.

"The identification of risk factors through electronic health systems allows health professionals to recognize and track them over time, helping both in clinical decision making, planning for outbreaks, and identifying transmission of diseases" [19]. For example, a "study of cancer patients allowed the tracking and analysis of diagnostic patterns, the number of investigations completed by physicians, and transfer of information as well as factors for the diagnoses" [20].

9 Conclusion

From our understanding of the limitations of the current HMIS and their various configurations, the modern HMIS system needs to be adaptable, transparent, and interoperable with other independent systems, like insurance systems and private health-care records. With several unknowns especially interfacing with other systems, the HMIS system that we propose will be appropriate and allows for any changes and modifications required to integrate properly to other systems.

Electronic digitized records are superior to traditional paper records because they decrease consolidation efforts, errors from handwritten records, and bulky physical storage requirements. EMR models present obvious advantages due to their ability to deliver a better record-keeping that tracks all medical interactions between the patient and the physician. It can also provide comprehensive health-care data across demographics and populations. Thus, a national health-care authority should envision a decentralized compendium of digital health-care data of individuals and populations as supporting data into error-free medical records, up-to-date health-care systems, and enhanced disease surveillance and monitoring.

Most EMRM initiatives are frequently government-initiated or -funded and hence broad-based to service the entire nation. Such initiatives run across national health-care systems to ensure medical care can be accessed by everyone and the data centralized by authorities. Unfortunately, if the patient switches from public care to private care and back again, their medical history becomes fractured, with parts of it centralized in different databases. However, our proposed personal EMR model aggregates their diverse records and allows the patient to selectively make them available to new or emergency providers for better data protection and privacy controls.

By readily integrating clinical applications into the system, modern HMIS reduces the burden of unnecessary cost and risk for hospitals faced with managing multiple health IT interfaces and ensures high degree of performance. Its standard-based technology is designed to support an exchange infrastructure that conforms to HL7 and SNOMED that facilitate sharing of health information in a secure manner.

As a modularized system, the digital-driven modern HMIS allows health-care organizations to reap real benefits from automated units, even as they continue to integrate other applications that contain data or support the patient care process, such as a hospital's laboratory information system and radiology information system. Leveraging best-of-breed, enterprise, web framework, HMIS allows for rapid development and flexibility and support software deployments at the shortest time possible. By readily integrating clinical applications into the system, modern HMIS reduces the burden of unnecessary cost and risk for hospitals faced with managing multiple health IT interfaces and ensure high degree of performance.

As more sophisticated HMIS progress continues, leading solution providers will converge and a preferred EMR model will emerge from full interpretational longitudinal base of EMRs. The ideal system would probably provide flexibility for patients and providers where medical records can be accessed across systems, and the interactivity of a permission-based EMR protocol would provide full patient control over their own health data. Ultimately, the technology should overcome the technical problems inherent in their solutions, like cybersecurity and privacy costs, on top of solving the bottlenecks from legacy systems. To reach large-scale adoption, HMIS should deliver lower costs and higher quality, while upgrading health-care delivery through patient autonomy and upholding professional ethics and ideals.

References

[1] Willis Towers Watson. (2020). Global Medical Trends Survey Report. Willis Towers Watson. 2019. Available from: https://www.willistowerswatson.com/-/media/WTW/Insights/2020/03/BPS-Survey-Report-2019.pdf

[2] Gunther, T. & Terry, N. (2005, Jan-Mar). The emergence of national electronic health record architectures in the United States and Australia: Models, costs, and questions. Journal of Medical Internet Research, 7(1), e3.

[3] DeMeijer, C. R. (2017). Blockchain in healthcare: Make the industry better. Finextra Research. 2017. Available from: https://www.finextra.com/blogposting/13801/blockchain-in-healthcare-make-the-industry-better.

[4] Ekblaw, A. & Azaria, A. (2016). MedRec: Medical data management on the blockchain. PubPub.2016. Available from: https://www.pubpub.org/pub/medrec.

[5] Heston, T. F. (2017). A case study of blockchain health care innovation. International Journal of Current Research, 9(11), 60587–60588.

[6] Angraal, S., Krumholz, H. M. & Schulz, W. L. (2017). Blockchain technology: Applications in Health care, Circulation: Cardiovascular Quality and Outcomes. arXiv 2017, arXiv:1706.03700.

[7] Zhang, J., Xue, N. & Huang, X. (2016). A secure system for pervasive social network-based healthcare. IEEE Access 2016, 4, 9239–9250.

[8] Rabah, K. (2017). Challenges and opportunities for blockchain-powered healthcare Systems: A review. Mara Research Journal of Medicine & Health, 1, 45–52.

[9] Esposito, C., Santis, A. D., Tortora, G., Chang, H. & Choo, K. K. R. (2018). Blockchain: A panacea for healthcare cloud-based data security and privacy? IEEE Cloud Computing, 5, 31–37.

[10] Raseena, M. & Harikrishnan, G. R. (2013). Secure sharing of personal health records in cloud computing using attribute-based broadcast encryption. International Journal of Scientific and Engineering Research, 1, 323–325.

[11] Omar, A. A., Rahman, M. S. & Kiyomoto, A. B. (2017). MediBchain – A blockchain based privacy preserving platform for healthcare data. In Proceedings of the International Conference on Security, Privacy and Anonymity in Computation, Communication and Storage, Guangzhou, China, 12–15 December 2017.

[12] Dwivedi, A. D., Srivastava, G., Dhar, S. & Singh, R. (2019). A decentralized privacy-preserving healthcare blockchain for IoT. Sensors, 19, 326.

[13] Zubaydi, H. D., Chong, Y. W., Ko, K., Hanshi, S. M. & Karuppayah, S. (2019). A Review on the role of blockchain technology in the healthcare domain. Special Issue Electronic Solutions for Artificial Intelligence Healthcare, 8(6), 679.

[14] Mohamed, H. & Ali, H. (2019). Blockchain, Fintech and Islamic Finance – Building the Future of the New Islamic Digital Economy. Boston/Berlin, De|G Press.

[15] Jiang, S., Cao, J., Wu, H., Yang, Y., Ma, M. & He, J. (2018). Blochie: A blockchain-based platform for healthcare information exchange. In Proceedings of the SMARTCOMP 2018: The 4th IEEE International Conference on Smart Computing, Sicily, Italy, 18–20 June 2018; pp. 49–56.

[16] Guo, R., Shi, H., Zhao, Q. & Zheng, D. (2018). Secure attribute-based signature scheme with multiple authorities for blockchain in electronic health records systems. IEEE Access, 6, 11676–11686.

[17] Liang, X., Zhao, J., Shetty, S., Liu, J. & Li, D. (2017). Integrating blockchain for data sharing and collaboration in mobile healthcare applications. In Proceedings of the PIMRC 2017: 28th Annual IEEE International Symposium on Personal, Indoor and Mobile Radio Communications, Montreal, QC, Canada, 8–13 October 2017.

[18] Dornan, L., Pinyopornpanish, K., Jiraporncharoen, W., Hashmi, A., Dejkriengkraikul, N. & Angkurawaranon, C. (2019). Utilisation of electronic health records for public health in Asia: A review of success factors and potential challenges. BioMed Research International, 2019, Article ID 7341841, 1–9.

[19] Chang, N., Dai, H., Jonnagaddala, J., Chen, C., Tsai, R. T. & Hsu, W. (2015). A context-aware approach for progression tracking of medical concepts in electronic medical records. Journal of Biomedical Informatics, 58, S150–S157.

[20] Lee, Y., Shin, S.-Y., Ahn, S.-M., Lee, J.-H. & Kim, W.-S. (2015). Validation for accuracy of cancer diagnosis in electronic medical records using a text mining method. Studies in Health Technology and Informatics, 216, 882.

Priyanka Jain, Ram Bhavsar, B. V. Pawar, N. K. Jain,
Hemant Darbari, Virendrakumar C. Bhavsar

Knowledge engine for a Hindi text-to-scene generation system

Abstract: We have developed Preksha – a text-to-scene generation system for the Hindi language. Hindi is a morphologically rich free word order language. Preksha has three major engines: a language engine, a knowledge engine, and a scene engine. The input text is processed by the language engine using a rule-based approach with a layered contextual pre-parser and a language parser based on the tree-adjoining grammar formalism. The language engine transforms the input text into parsed derived trees represented in a language-independent Extensible Markup Language format. After a review of our earlier version of the language engine, this chapter mainly focuses on the knowledge engine of Preksha that takes these parsed trees as its input and generates output for further processing by the scene engine. The knowledge engine consists of a knowledge extraction module and a knowledge representation module. We introduce object visualization features which support these knowledge engine processes. An example of a scene having multiple objects with spatial relations generated using the knowledge engine is given.

Keywords: tree-adjoining grammar, knowledge extraction, knowledge representation, automatic text visualization, natural language processing, text-to-scene conversion

1 Introduction

This chapter presents a text-to-scene generation process for the purpose of automatic text visualization (ATV). This research focuses on natural language comprehension of a given Hindi language input text and transformation of it into a structured knowledge form. This work is named as Preksha, the only known ATV work intended for language Hindi (i.e., an Indian language). The ATV is a relatively new interdisciplinary area under artificial intelligence field. It is challenging because it is not easy to automatically generate a scene which exactly corresponds to the user's mental image and its expected

Priyanka Jain, N. K. Jain, Hemant Darbari, Centre for Development of Advanced Computing, New Delhi, India, e-mail: priyankaj@cdac.in, nkjain@cdac.in, darbari@cdac.in
Ram Bhavsar, B. V. Pawar, North Maharashtra University, Jalgaon, Maharashtra, India, e-mail: rpbhavsar@nmu.ac.in, bvpawar@nmu.ac.in
Virendrakumar C. Bhavsar, Univeristy of New Brunswick, Fredericton, Canada, e-mail: bhavsar@unb.ca

https://doi.org/10.1515/9783110713633-012

visualization. It is especially more complex due to the morphologically rich and free word order nature of the Hindi, that is, an Indian language. The issues related to Hindi language processing and adopted suitable solutions are discussed in the subsequent sections. Section 2 is dedicated to provide necessary background work of Preksha that is essential to understand before proceeding to the knowledge engine. It discusses the Preksha Systems challenges and preparation like resource organization, degree of visualization (DoV), object visualization features (OVFs), and language processing in Preksha work. Section 3 describes the knowledge engineering which consists of knowledge extraction (KEE) and knowledge representation (KER). Section 4 shows Preksha's KEE which focuses on extraction mechanism, object selection, and context-based background image retrieval. Section 5 discusses KER for preparing scene knowledge, and xml preparation with Java object marshaling. The result is presented and discussed in Section 6. Finally, Section 7 concludes the chapter by exploring its contribution.

1.1 Preksha

Though this chapter is an extended dimension of our previously reported works [12–21, 48–49, 52–54], an incremental further work with deeper level of dimensions is contributed.

1.2 Challenges

In this section, we explore the challenges in the current research problem and formulate the problem statement. A common issue in all ATV systems is that a single instance of text clause (TC) can be visualized semantically in multiple forms, be it computer automated visualization or human mental imagination. The mental imagery from the input language is based on one's language comprehension, imagination power, and personal experience. It causes many-to-many mappings between input text and generated scene output. Well, vice versa, a single natural language text segment can convey the similar semantic by various syntactic constructions of utterances using different word group orderings. This is not a specific phenomenon to any particular language.

Generally, the languages are ambiguous by nature and it is more specific when it comes to the consideration of visualization. For a knowledge-based spatial reasoning, the focused problems are mentioned as under-specification, vagueness, uncertainty, context, and frame of reference as discussed by Tappan [26]. The problem is severe for Hindi language processing, where the main complexities are the linguistic divergence, free word order, semantic ambiguity, resource deficiency, automation complexity, and other operational challenges. Hindi language is in the family of Indian languages and has MoR-FWO nature. It has complex morphology with varying degrees of richness [2]. The main verb is normally in the terminating position of the sentence in the Hindi language. Therefore, the remaining sentence constituents occur in any position in some respect. Recently, an automatic title generation tool for Hindi short stories [50] is proposed to work with highest priority nouns in the given input to get simpler solution for MoR-FWO syntactic constructions.

1.3 Syntactic variations

In this way, it gives a wide range of possibilities to generate or visualize a scene out of Hindi text understanding. Let us consider a text example "The towel is on the table." This is a short and simple TC with a single verb, two nouns, and one preposition. For equivalent example in the Hindi language, the syntactic variations (SV) are analyzed as SV = टेबल पर तौलिया है, तौलिया टेबल पर है, टेबल पर है तौलिया, तौलिया है टेबल पर. Therefore, |SV| = 4. The morphological variations (MV) of sentence constitutes are MVtowel = टॉवल, तौलिया, अंगोछा. Here, MV-table = टेबल, टेबिल, मेज़, पटल, therefore, |MV-table| = 4; MV-on = पर, के ऊपर, therefore, |MV-on| = 2; MV-is = है, रखा है, रखा हुआ है, therefore, |MV-is| = 3. By taking all variations into account, the possible count of Hindi linguistic syntactic constructions can be calculated as follows:

$$|SV| * |MV\text{-}towel| * |MV\text{-}table| * |MV\text{-}on| * |MV\text{-}is| = 4*3*4*2*3 = 288$$

There are 288 possible constructions of Hindi language utterance, which can be imagined by a user or ATV system in multiple scene forms. Figure 1 provides some self-clicked in-house images to understand the many-to-many mappings in the text-to-scene conversion system. For the given text example, "The towel is on the table," any of these types of pictures can be chosen or prepared by an ATV system. In this way, a solution is needed to transfer the information from multiple syntactic constructions to a suitable visualization.

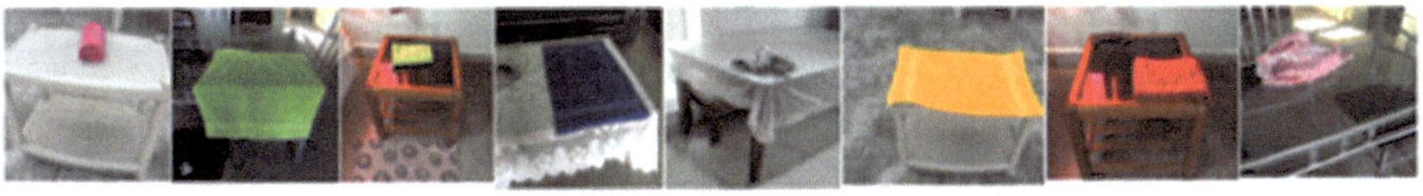

Figure 1: Some of the possible visualizations corresponding to sample text.

We propose KEE and KER as important key components to bridge the gap in between language comprehension and scene engine. With this objective, problem statement focuses on to formulate a knowledge engine with:

- **KEE** from semantically and contextually related multiple sentences in Hindi language text.
- **KER** in a form of discourse should be factual, adequate, language-independent, and technology-independent.

To support the proposed design, we define the scope of this research for simple and compound Hindi language sentences. These sample sentences should have a reference of physical objects which can be visualized, that is, can be seen by human eyes. Looking into the linguistic complexity of considered input language, we investigate rule-based approach which will be extendible for the further scope of research enhancement.

1.4 Spatial ambiguity

The spatial ambiguity [12, 19] and argument scrambling [3] are major issues in Hindi language processing which is handled in the prior natural language processing (NLP) engine of Preksha. However, parsing a complex Hindi language is not in the scope of this research work. This chapter focuses on Hindi knowledge engine, a later component that postprocesses the output of language parsing. This chapter proposes OVFs set in Section 1.4.3, to impose structure on lexical information, both at the object (lexical items) and meta (lexical concepts) levels of the lexicon. OVFs help in structuring document-level event descriptive information for KER. This structured discourse knowledge is the basis of the scene planning and consequently scene synthesis.

Spatial relation as postpositions in Hindi language is highly ambiguous. They have different semantic roles depending on the arguments they select and the sites they get attached to. Sense selection is the identification of postposition, as it plays a role in spatial relationship in the scene. The Hindi postposition "से" takes two equivalents in the English language as "from" (ablative) and "with" (instrumental). Consider two different sentences with this postposition as "सीता ने चम्मच से खाया," that is, Sita ate with a spoon and "सीता से खाया नहीं जाता," that is, Sita cannot eat. The first sentence is visualizable, whereas the second sentence is difficult to visualize. The example of another postposition "को" with English equivalent "to" (accusative) and "for" (dative) as – "राम को पुस्तक दी," that is, "The book is given to Ram" and "राम को बुखार है," that is, "Ram has a fever." Another example with postposition is "पर," which is basically a locative modifier. The corresponding sentences for locative sense are "दीवार पर घडी है," that is, "The clock is on the wall." It takes a different sense in sentences like "भगवान् पर श्रद्धा है," that is, "Have faith on God" and "ट्रैन समय पर है," that is, train is on time. In addition, the adverbial phrases in the Hindi language also take help of postposition in conveying the meaning like "जल्दी से," that is, "hurriedly," "पास में," that is, nearby and "तेज धुप में," that is, "in hot sunlight." The issue cannot be resolved in isolation with modifier only. The role of a modified head noun is very important to generate an appropriate scene. From a computational perspective in cognitive science, two main approaches toward grounding spatial relations [42] are the spatial template theory and the constrained connectionist approach. The connectionist approach is data-driven, whereas the spatial template theory favors the rule-based approach which influences the sense disambiguation in Preksha system.

2 Overview

Before proceeding to the related work, we present the basic process flow and internal components of the Preksha system in Figure 2. The basic architecture for proposed Hindi text-to-scene conversion is presented in [17]. It has three major engines: language processing engine, knowledge engine, and scene engine. This is to mention here that having a rule-based approach, Preksha research is well supported by an offline linguistic resource repository named Preksha-R.

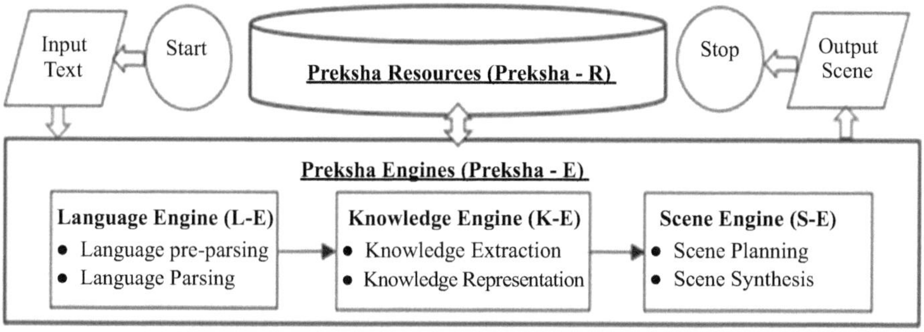

Figure 2: Process flow of an ATV system.

The input text submitted by the user is grammatically processed in language engine (LE) for language understanding. The prepositions in the English language are postpositions in Hindi and other Indian languages. The ambiguity and challenges in working with these postpositions are explained in [12]. For Hindi language processing, we used our rule-based approach [17]. We have implemented a layered contextual pre-parser and language parser [13] based on tree-adjoining grammar (TAG) formalism [22, 27]. The implementation is based on the Earley's algorithm [8]. It transforms this language information into a language-independent knowledge structure in Extensible Markup Language (XML) format [29], as mentioned in our prior work [14]. The spatial intelligence work for the Hindi language is discussed in [19]. The system processes the extracted knowledge for scene planning in Virtual Reality Modeling Language (VRML) file format [26] as mentioned in our previous work [16]. Using scene grammar by extracted knowledge, a scene is planned with background preparation, objects selection, spatial relation, merging and positioning, scene placement, and final image rendering. A three-dimensional (3D) scene synthesis is done from VRML file using Java3D [31] APIs in the scene processing engine. The SceneGraph [35], a data structure in computer graphics, is used to arrange the visible components of a scene by maintaining a hierarchy of related objects in a scene. The initial evaluation strategy for Preksha research is discussed in [15]. A reverse work, knowledge acquisition from an input image and generate language Hindi, is explained in [21]. The utility of this approach is useful in cognitive support and language learning, as discussed in [18, 20].

Though the focus of this chapter is on knowledge processing engine, the presented literature survey in this section discusses about the related work of the overall ATV system also. Readers may refer to a detailed survey presented in [10] on varied parameters for the work on text-to-scene generation systems. It discusses the requirements and challenges for developing systems that are capable of visualizing descriptions expressed in a natural language. However, we discuss in short some important applications for readers' reference in this chapter. The very early works are Adorni et al. [1] and Winograd [28] with initial concepts and basic works aimed at recreating static 2D scenes. The Automatic Visualization of Descriptive Text [36] processes the information with a directed graph named Parts of Spatial Information for an Natural Language Visualization (NLV) application. The work in Ma [24] offers a lexical knowledge through a generative lexicon, which can be used for disambiguation and commonsense inference to fill unspecified argument structures. Lu and Zhang [23] use a detailed knowledge base for reasoning for the automatic interpretation of the text and for generating the layout of a scene. The WordsEye System [6] generates dependency representation to produce a description of the arrangement of objects in a scene. Scenario-Based Lexical Knowledge Resource is a knowledge base to represent lexical and commonsense knowledge [43] as a core of this system. Finally, an image is formed from this scene description. A richer and more expressive vocabulary description is done for lexical information by Pustejovsky and Boguraev [25]. The knowledge-based spatial reasoning is presented by Tappan [26] with lexical semantics theory. A method in [11] presents the semantic representation of natural language expressions (texts, sentences, phrases, etc.). The SDAS system [34] consists of three modules: story understanding module, stage directing module, and action generating module tightly coupled with their knowledge bases for simple children's stories in Japanese.

The major work for ATV is reported in [4–7, 30] in English and other languages. As per the best knowledge of author, there is no research found for Hindi or other Indian languages except our work in [17]. Though the other presented works are not available to use, extension, or customization for Indian languages, we present an approach which is language independent and extensible to other Indian languages also.

2.1 Resources

In Preksha-R, the lexicon repository is enriched with annotated lexicons with their meanings, synonyms, grammatical categories, and OVFs. This repository also maps the noun objects with 3D object model's name for scene planning and synthesis process. Apart from annotated lexicons, Preksha-R upholds the grammatical rules and linguistic heuristics for computational components of NLP engine, namely, pre-parsing and parsing. The Preksha-R is utilized throughout the processing of language, knowledge, and scene engine.

2.2 Degree of visualization (DoV)

This is to note here that the text can be visualized well if it has references of physical noun objects which can be seen by human eyes. But it is not easy for abstract nouns to convert into a scene formation. This is the problem of synthesizing a scene corresponding to text conveying feelings, abstract nouns, and some verbs (like action or emotion). The examples of some abstract nouns in Hindi are “अहंकार,” that is “ego,” “विचार,” that is, “opinion,” “ठंडी,” that is, “cold,” “ताजी,” and “fresh,” “खुशबू,” that is, “smell,” and “ऊर्जा,” that is, energy, which are difficult to visualize apparently. It is difficult for a synthesized scene to have accurate representation corresponding to such type of information. These feelings can be understood by a human with some references but not easy for an ATV system to convey it with full fidelity. Consider an example in Hindi language “ठंडी हवा में ताज़ी खुशबू है” with corresponding English text “Fresh smell is in the cold breeze”; the comprehension of this sentence to a human gives more emphasis on feeling instead of a mental image. Even, it is very difficult for a human being to visualize the scene corresponding to this text, which is syntactically and semantically correct a simple sentence. This is the limitation of any ATV system irrespective of language considered. To simplify this issue in ATV, we classify the input text as based on its DoV. A text is considered as visualizable in Preksha system if its DoV ≥ 1. This is a count of the visualizable noun objects in the input text. To calculate the number of visualizable noun objects in an input text, we discuss OVFs in the next section.

2.3 Object visualization features (OVFs)

Before proceeding further, we discuss some of the sematic knowledge bases here. The WordNet [42] is commonsense knowledge as a lexical taxonomy with 101,863 nouns that provides using hypernyms, hyponyms, holonyms, and meronyms. An extension of FrameNet, VigNets [40, 44], consists of a set of intermediate frames (vignettes). This is knowledge to bridge the semantic gap between semantic frames of FrameNet and low-level graphical frames. Nevertheless, none of these knowledge bases provide information regarding physical attributes of object categories for visualization purpose. The situation is worse for Indian languages. We introduce OVFs similar to grammatical semantic features [9], to calculate DoV and further processing of information. The grammatical semantic features are based on linguistic theories, whereas OVFs are extended on the aspect of visualization of physical objects as a sentence constituent. The OVFs are abstract arguments, designed to tag the noun entities mainly. In general, the OVF sets are oriented for linguistic endeavors and visualizability for knowledge management. It consists of roles like “supportability” and “visualizability” of the noun objects in a scene to be synthesized. It also takes care of default attributes like shape, color, weight, size, and supporting ambiance of visualizable physical objects. The OVFs in Preksha help in normalizing the input text for minimal structure ambiguity in syntactic parsing also. The visual feature set for "noun" objects with its category, feature, ID, and description is given in Table 1.

Table 1: Object visualization features set.

Feature	ID	Sample values
Visualizability		
Is it visualizable	VI	0 = no, 1 = yes
Is it supportable	SU	0 = NA, 1 = supporter-to, 2 = Supported-by
Is it background	BG	0 = no, 1 = indoor, 2 = outdoor
Size dimensions	SD	(X, Y, Z)
Shape	SP	Shape-ID
Color	CL	Color-ID
Texture	TX	0 = any, 1 = wooden, 2 = metal, 3 = brick, 4 = plastic, etc.
Default background	DB	0 = NA, 1 = room, 2 = ground, 3 = jungle, 4 = sky, 5 = hospital, 6 = kitchen, 7 = office, 8 = washroom, etc.
Default supporter	DS	0 = NA, 1 = table, 2 = floor, 3 = tree, 4 = plate, etc.
Semantic		
Is it temporal	TM	0 = no, 1 = yes
Gender	GN	0 = NA, 1 = male, 2 = female, 3 = neuter
Named entity	NE	0 = NA, 1 = person, 2 = organization, 3 = location, 4 = number, 5 = object
Animacy	AN	0 = rest, 1 = human 2 = animate
Quality	QT	0 = NA, 1 = good, 2 = bad

The feature values of noun entries are annotated according to 3D models and stored in Preksha repository, as mentioned in Section 1.4.1. Sample feature sets of "flowerpots" as an object and "room" as a background is shown here.

2.3.1 Visualizability feature set

VI, SU, BG, SD, CL, TX, DB, DS, TM, GN, AN, QT

Object "Flower-Pot"

1, 1, 0, (1.7, 1.1, 1.7), Prop-0001C, Prop-CLR-GRN, 2, 1, 1, 0, 0, 5, 0, 1

Background "Room"

1, 1, 1, (0, 0, 0), 0, Prop-CLR-GRY, 3, 0, 0, 0, 0, 0, 0, 1

It represents that the mentioned 3D model for object "flowerpot" is visualizable, supportable, has 3D model properties (dimensions, color, shape, and texture), with default supporter as "Table" and default background as "Room." It also contains the semantic features like "temporal," "gender," "named entity," "animacy," and "quality." All the noun lexicons are annotated with OVFs and are stored in Preksha-R for use in all forthcoming processing.

3 Language engine

The detailed work on Preksha language processing engine is mentioned in our prior work [13, 17]. For a quick look at language processing in Preksha, consider Hindi language along with the corresponding English language equivalent.

Hindi text (HT)	एक लकड़ी की कुर्सी और गोल टेबल धूप में है. गोल टेबल पर बैगनी तौलिया के ऊपर हरा फूलदान रखा है और उसके पास एक फूल सुबह से पड़ा है।
English text (ET)	There is a wooden chair and round table in the sunlight. A green flowerpot is placed above the purple towel on a round table and a flower is lying near the green flowerpot since morning.
Text clouse (TC_a)	एक-लकड़ी-की/JJ कुर्सी/NN और/CC गोल/JJ टेबल/NN धूप/NN में/PSP है/VM
Text clouse (TC_b)	गोल/JJ टेबल/NN पर/PSP बैगनी/JJ तौलिया/NN के-ऊपर/PSP हरा/JJ फूलदान/NN रखा-है/VM
Text clouse (TC_c)	उस/NN के-पास/PSP एक/JJ फूल/NN सुबह-से/RB पड़ा-है/VM

Here, the visualizability in the input text is the total number of visualizable objects which are six; these are धूप "sunlight," कुर्सी "chair," टेबल "able," तौलिया "towel," फूलदान "flowerpot," and फूल "flower." The DoV of the input text is six which is greater than one; therefore, the input text is considered as visualizable in Preksha. This input text has two sentences: one is a simple sentence and second is a compound sentence. Since the second sentence has a conjunctional connection between two sentential clauses, it can be easily divided into two TCs. We divide the second sentence into two simpler structures for easy and unambiguous text parsing process. This approach for simplifying construction is also suggested in [33] as it avoids the unsatisfying performance because of long sentence processing. In the given example, we have labeled the semantic roles for the simplified clauses TC_a, TC_b, and TC_c. It presents the pre-parsed output of these three clauses from LE's pre-parser component. The pre-parsing of a text consists of a morphological analyzer, POS tagger, chunking, and relocation processes. One example of the pre-parser output is shown.

For TC_c, the token "उसके पास" is processed into 'उस'/NN 'के-पास'/ PSP using repository Preksha-R and pre-parsing components. Our language parser is based on TAG formalism. The elementary building blocks used in the TAG are structured tree objects. The TAG produces derivation and dependency trees after operations, which are important both syntactically and semantically. Based on our work as in [13], a postposition spatial relation-based TAG grammar is formed to get a very optimum controlled parsed structure of Hindi language.

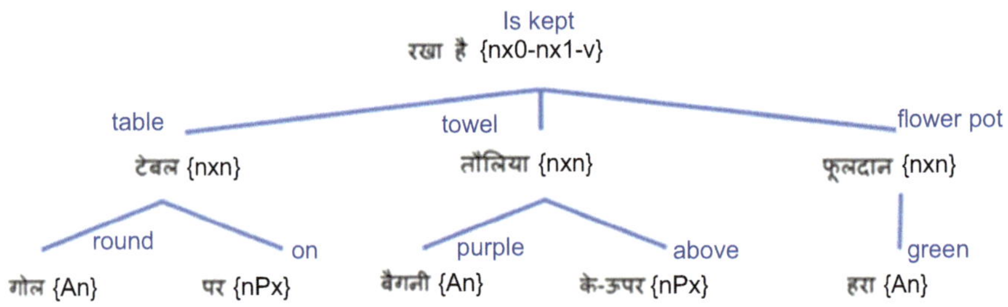

(a). TAC Parsed derivation Tree for TCb

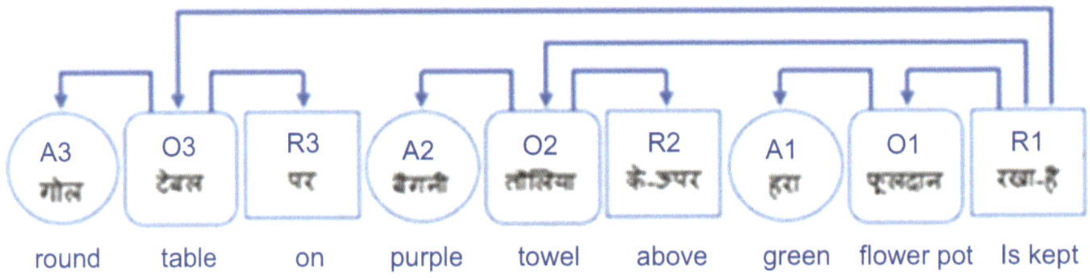

(b). TAC Parsed derived relation for TCb

Figure 3: NLP engine outputs for TC_b.

This grammar leaves minimum ambiguity and generates correct specific structures. A derivation tree is a dependency relation of the TAG trees as per their attachment to root tree. A derived tree is a semantic relation graph between the lexicons of sentence constituents. A TAG parsed tree corresponding to TC_b is presented in Figure 3(a). Figure 3(b) shows dependency relations simulated through TAG-derived tree. Here, "O" refers to noun "Object" model to be visualized in the final scene. The "A" is an associated attribute with "O," and the "R" is the relation between objects.

Constructing a discourse knowledge needs to relate the pronouns with its referential entities in the multiple sentences. The process of binding the referring expression to the correct antecedent, in the discourse, is called anaphora resolution or pronominal resolution. We have handled anaphora resolution at limited scope due to argument scrambling issue. The anaphora resolution work for Hindi language is presented in [37, 39] which are focused on resolving pronoun entities. The different syntactic behaviors for certain pronominal forms like reflexive, locative, relative, and personal pronouns can be resolved quite successfully using some specific rules with the dependency information. With the help of an exhaustive list of pronoun categories, the rule-based module uses the dependency relations and other information for anaphora resolution. It is observed that the reflexive pronouns (अपना, अपनी, अपने) and nonpossessive reflexive pronoun (अपने-आप self, स्वयं myself) have no significant visualizability as they are representing oneself. The place pronouns are locative (यहां here, वहां there) and refer to location or place. In this case, we take the dependency reference of a locative noun, based on semantic features TM, GN, NE, AN, QT as mentioned in Table 1. The relative pronouns जो, जिसको, जिसने, and so on are handled with a semantic feature role of "supporter" and "supported-by." These are manipulated using the Gazetteer method [40] for recency and animacy in anaphora resolution. In case of further ambiguity, identification of the actual position is done using role matching. An example from sample text is presented here.

Sentence two of Hindi text (HT)	गोल टेबल पर बैगनी तौलिया के ऊपर हरा फूलदान रखा है और उसके पास एक फूल सुबह से पड़ा है। A green flowerpot is placed above the purple towel on a round table and a flower is lying near the green flowerpot since morning.
Sentence decomposition	(TC_b) गोल टेबल पर बैगनी तौलिया के ऊपर हरा फूलदान रखा है (TC_c) उस (हरा फूलदान) के पास एक फूल सुबह से पड़ा है।
Sentence reference	(TC_b) गोल टेबल पर बैगनी तौलिया के ऊपर हरा फूलदान रखा है (TC_c) हरे फूलदान के पास एक फूल सुबह से पड़ा है।

4 Knowledge engine

Visual knowledge consists of visual dimensions which are useful for data encoding. There are seven parameters presented by Bertin [38] for visual information, namely, position, shape, size, value, orientation, hue, and grain. Accordingly, major measures are taken care of position, shape, size, color, and rotation while working with Preksha system. As the discourse knowledge [41] is a language talking about language, the objective of Preksha knowledge engines is to obtain the information from NLP engine and to represent it in a language-independent abstract structure. This representative abstract structure should be able to generate the corresponding scene for visualization.

This section talks about the quality measures taken into account before the design of Preksha knowledge engine process. The knowledge engine processes for converting language to a logical form for encoding knowledge, belief, action, feeling, goal, desire, preference, and all other mental systems in artificial systems. It is designed to be unambiguous with precision to the level of detail necessary. The naturalness is the criteria of easiness by which it can be interpreted by common people. The inferential adequacy of Preksha knowledge base is a well-used representation to infer what it needs to. The representational efficiency is to do its reasoning in terms of time or computational resources, as appropriate. The extensibility in terms of technology and scope is another important feature of a Preksha knowledge base. In Preksha, we deal with explicit and implicit knowledge from an input text. The design of the knowledge base covers the aspects of compactness, the sufficiency of information, language independence, and expandability. To deal with these aspects, the Preksha knowledge engine is designed into major components as KEE and KER. The overall flow of Preksha knowledge engine is shown in Figure 4.

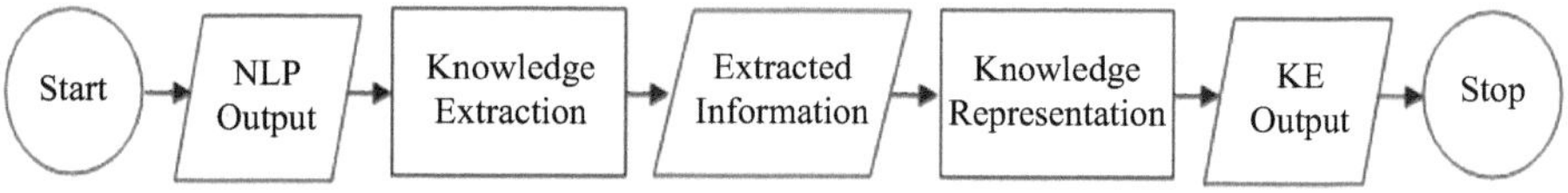

Figure 4: Knowledge engine.

The parsed derived tree from NLP engine in XML form is the input of knowledge engine. It is further processed to extract knowledge using DoV and OVFs. The KEE engine reads the elements of this parsed derivation XML. A novel approach for tree traversal is used, considering the concept of postpositions [12] in the Hindi language. With the help of position dependency value, the extracted knowledge gets processed for KER into another XML format, that is, the output of knowledge engine. While processing, it holds the extracted information in a Java class objects. The XML processing is done using JAXB API [33]. It provides methods for unmarshaling (reading) which transforms XML instance into Java content trees, and then marshaling (writing) which converts Java content trees back into XML instance documents. A qualitative model for the representation of spatial knowledge is developed.

A 3D coordinate system defines an origin, orientation (major axis), direction (up direction), and scale. This is discussed in previous sections that the text-to-scene conversion is mainly about rendering physical noun objects into the virtual world according to their attributes and spatial relations extracted from the input text. Semantic spatial concepts are based on comparisons between perceived magnitudes of the objects. As mentioned, OVFs take care of visualizable attribute (mainly adjective) of 3D models corresponding to noun objects. The spatial relation defines the placement strategy so that all the objects get an appropriate position in the synthesized scene. To describe the locations of the objects, a 3D space is used which provides six-direction coordinate system. The directions are left, right, up, down, back, and forward with representation as śX, śY, and śZ, respectively. The Preksha system takes care of 51 reference Hindi spatial relations. The relations are classified in symbolic notations as R1 ↓ (नीचे Down), R2 ↑ (ऊपर Upon), R3 ← (बाएं Left), R4 → (दाएं Right), R5 ↔ (पास Nearby), R6 ←→ (दूर Far), R7 < (छोटा Smaller), R8 > (बड़ा Bigger), R9 # (भीतर Inside), and R10 @ (बाहर outside).

Having a free word order language, the sequence of placing objects can conflict due to sentence structure ambiguity. This is illustrated in following four example sentences – four sample clauses गोल टेबल पर बैंगनी तौलिया है/बैंगनी तौलिया गोल टेबल पर है for "towel on table," गोल टेबल बैंगनी तौलिया के नीचे है/बैंगनी तौलिया के नीचे गोल टेबल है for "table down below the towel." The knowledge engine is efficient to understand the reverse spatial relations in the syntactic variants in Hindi language. The OAR model [32] matches the process of human cognition and presents a knowledge manipulation by objects "O," attributes "A," and/or relations "R," among these. Considering this, the first two sample clauses can be interpreted as [(A1)O1] [(↑)R2] [(A2)O2] and second two clauses can be represented as [(A2)O2] [(↓)R1] [(A1)O1] as shown in Figure 5. Here, "O" and "A" are the symbolic representations of object and its attribute correspondingly. "R1" and "R2" are the mentioned relation between the two objects "A1" and "A2."

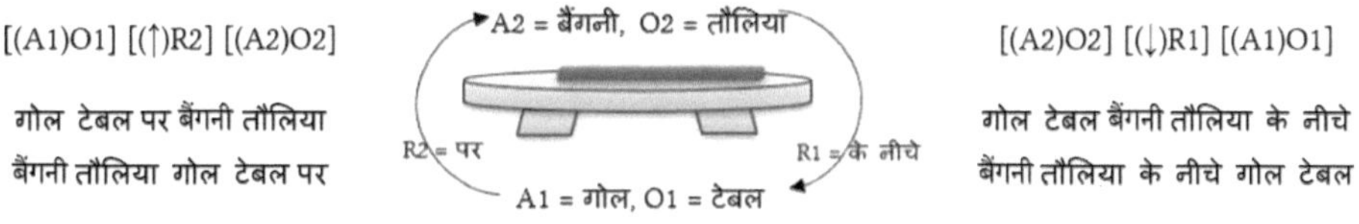

Figure 5: Reverse spatial relations.

5 Knowledge extraction

The knowledge engineering is the process of extraction of semantically structured information from preexisting parsed text. The extracted knowledge from input text is a collection of TC. Therefore, KB = TC1, TC2, TC3, TCn, where "n" is the number of logical TCs in the text. A TC is a collection of local word groups (LWG), TC = LWG1, LWG2, LWG3, ..., LWGm where "m" is the number of word groups in a TC. The LWG consists of one or more semantically related words or entities. An entity is a three-touple information corresponding to physical objects, that is, ENT = (O, A, R). In order to extract entities, attributes, and/or relations, we use a triplet in a sentence as a relation "R" with object "O" and the attribute "A" (OAR). We extract and structure the knowledge on the basis of this triplet.

5.1 Processing the parsed tee

The KEE is done by processing a syntactic parsed derived tree, which is a directed rooted tree. The parsed tree is a syntactic representation of the close linguistic relation in between the lexical tokens of input text (TC_b from Figure 3) as shown in Figure 6. It is rooted with symbol "S," which represents the "Sentence" as shown in Figure 6(b). Mainly, a sentence consists of two parts: subject and predicate. In Preksha TAG grammar, we considered these parts as "NP" (noun phrase) node and "VP" (verb phrase) node. The further child nodes are derived from these nodes, based on grammar and linguistic theories. A parent node of two or more leaf node siblings covers an LWG. Gradually moving upward, it covers a broader group of long-distance dependencies. To extract the required relations from a derived tree, a new traversing algorithm is implemented. It traverses derived tree in "right-root-left" manner and identifies dependency among the constituents. The analysis starts at the right most nodes where the verb holds the terminating position. It further moves to extract information from dependent clauses.

It helps in identifying the three information, namely, (a) "the relation of the object," (b) "the object on which this relation is applied," and (c) "attribute of the object in a close proximity." As discussed in our work [12, 19], spatial relations are mainly postpositions in the case of Hindi language. A "postposition or relation" always follows its heading noun; therefore, its attachment position is the right side of the head noun. An "adjective or attribute" is a succeeding token to the head noun; therefore, its position of attachment is left side of the head noun. The intra-group dependency of TC_b. as shown in Figure 6(a), this clause example "गोल टेबल पर" (on round table) takes the right-root-left traversal and identifies "पर" (above) as spatial relation for preceded noun group, "टेबल" (table) as the head noun object and "गोल" (round) as an attribute of noun object "टेबल" (table). The arrows indicate semantic dependencies between the tokens.

The intergroup relations also help to identify the role "Supporter-to" in the scene. Here, the टेबल (table) is "Supporter-to" तौलिया (towel) and तौलिया (towel) is "Supporter-to" फूलदान (flowerpot). The sentence constituent हरा फूलदान (green flowerpot) in TC_b matches with its reference in TC_c "उस." The next section discusses on anaphora resolution of this reference with its morphological oblique form हरे फूलदान (green flowerpot). A data structure is prepared in class object "EntInfo" by parsed derived tree for each noun objects and related information. The interrelation of the semantically related clauses of the sentence is identified using Unique-ID.

This is to note here that the presented KEE engine is implemented to work with Preksha TAG grammar, which is very much focused on Hindi linguistic etymology. However, the approach of engine processing is extendible for any other Indian languages also. The intra-word group traversing is applicable to languages like Urdu, Punjabi, and Gujrati with post-positional grammars. In a similar manner, the inter-word group traversing is valid for languages with subject–object–verb nature. Therefore, this approach for information analysis supports the extendibility for languages mainly which falls under free word order nature.

6 Extracting the 3D objects

The Preksha-R maintains the annotation of various 3D object models mapped with multiple lexicon names along with their attributes. The knowledge engine is responsible for selecting an appropriate 3D object model from the repository for visualization purpose. Here, we take an example of lexicon token 'लकड़ी-की लाल छोटी कुर्सी' (a wooden red-colored small table) in some input text. Figure 8 shows the sample annotated object 3D models in Preksha-R.

According to the input lexicon, the expected 3D object model should have three attributes: (a) "लकड़ी-की" means the texture is wooden, (b) "लाल" for color "red," and

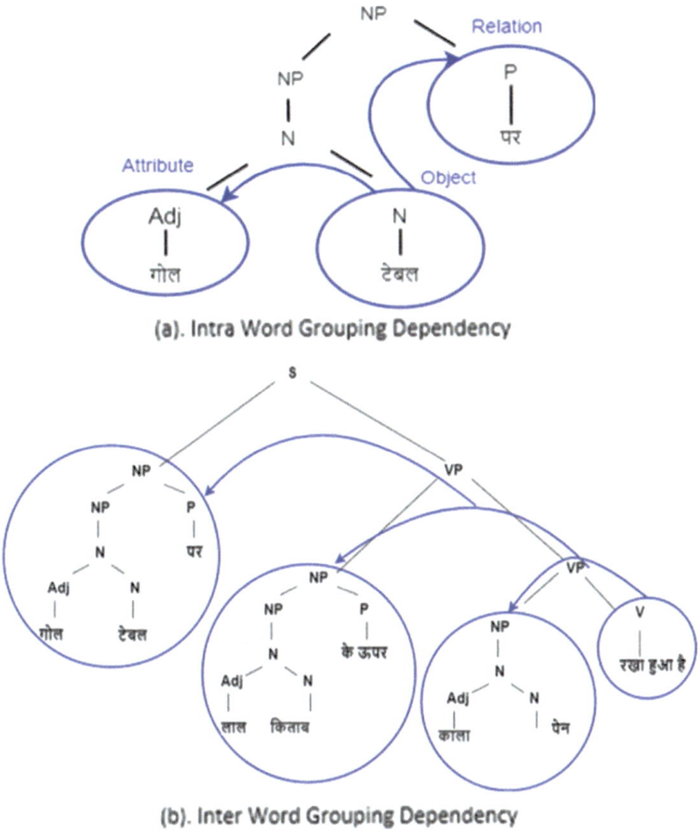

Figure 6: Local group dependencies for TC_b in Figure 3.

(c) "छोटी" with size "small." Suppose that the available entries in Preksha-R are as shown in Table 2 for selection of 3D object model for "chair."

Table 2: Lexicon and object mapping in resource repository.

Annotated tag name	Attribute	Default background	File name
सोफ़ा, सोफा, कुर्सी	Red, white, big	Drawing room, home, hotel, lounge	**Sofachair**
सोफ़ा, सोफा, कुर्सी	White, small, plastic	School, playground, home, room	**Kidchair**
कुर्सी, बेबी-कुर्सी, बेबी-चेयर	Gray, rotating	Study room, office, home	**Office_ch1**
कुर्सी, ऑफिस-कुर्सी, स्टडी-चेयर	White, big, wooden	Home, room, park	**Chair_white**
कुर्सी	Blue, wooden	Home, room	**Bluewood1**
सोफ़ा, सोफा, कुर्सी	Small, red, stool	Home, room, bar	**stoolred**
कुर्सी, स्टूल	Green, metal	Home, room, office	**Green_chair**

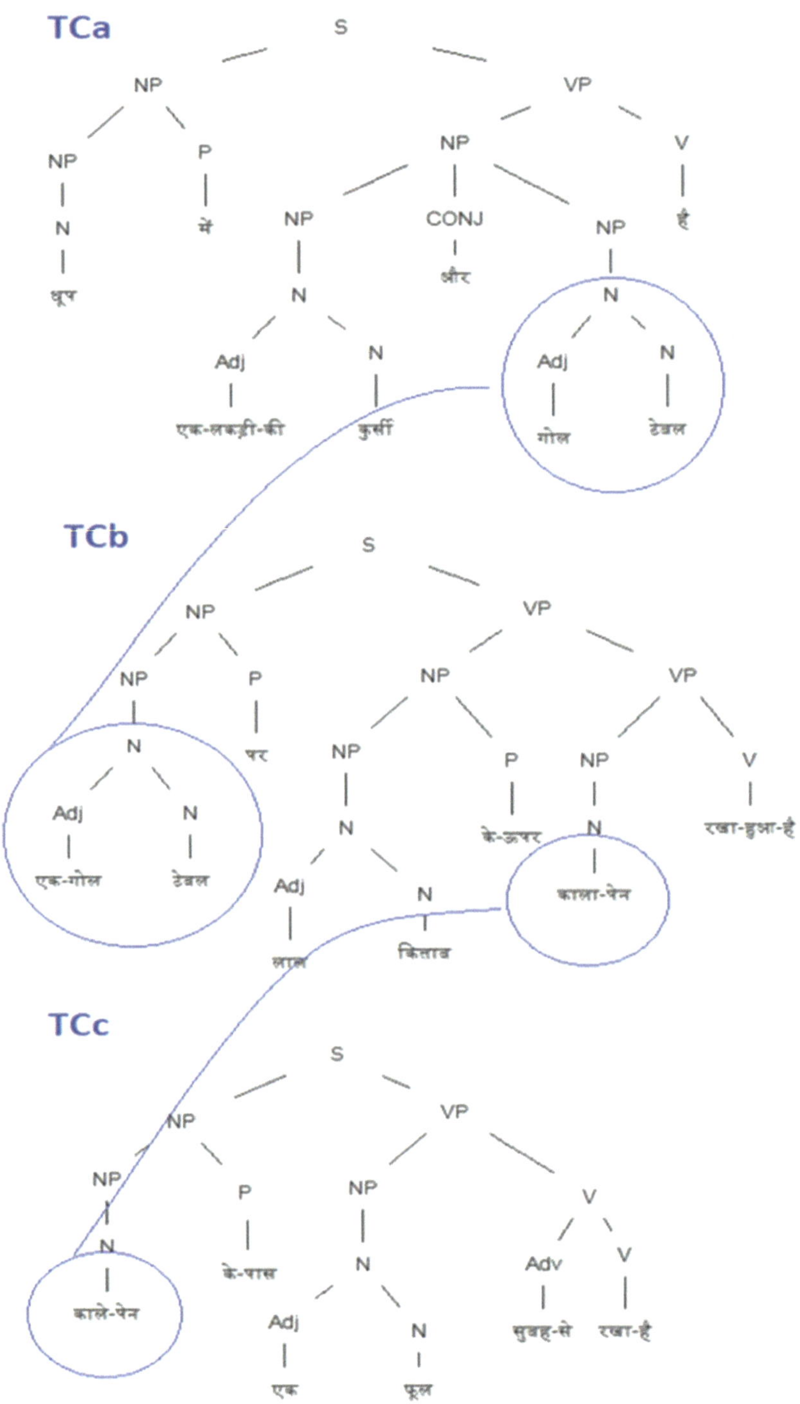

Figure 7: Inter-clausal relations for input text.

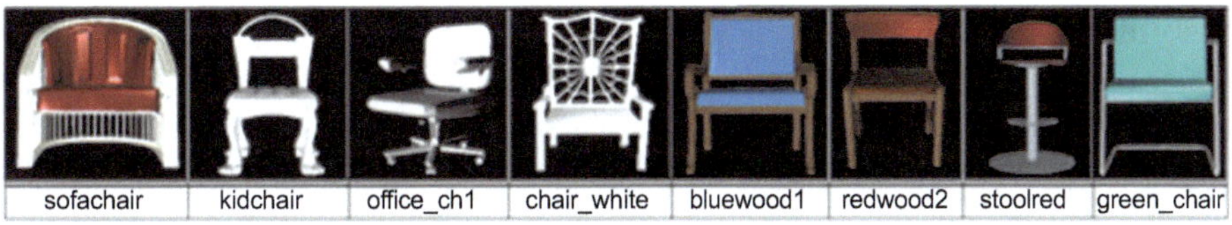

Figure 8: Appropriate background selection.

According to Table 2, there are three chair models with reference to the wooden attribute, three references to red color, two references to a small table. Only one single 3D object model "Redwood2.wrl" is mapped with all three required attributes, that is, "लकड़ी-की लाल छोटी कुर्सी" with the wooden texture, small size, and green color.

In the absence of availability of object model with all required attributes, it selects the object model which has maximum numbers of attributes matched for scene planning. In the case of retrieval of multiple objects, the system uses the first occurrence of 3D object model. The unavailability of an appropriate object in Preksha repository is taken care by contextual analysis of input text. The default background "BD" parameter of OVFs helps in analyzing the selection of appropriate objects of the scene to be synthesized.

7 Planning the background

Context-based background selection is very important for appropriate scene generation process. The system processes the explicit and implicit knowledge of background information in input text. The information of background may be mentioned explicit in input text, like "कमरे में," or "मैदान में." Elsewhere in the absence of explicit information, we need to extract the default background from references of other objects. For example, a "refrigerator" indicates the default background as "kitchen" and a "football" indicates "playground." Whereas it is not necessary, still it gives an indication to produce an implicit knowledge.

Figure 9: Example of annotated background images.

If a text has multiple references of various background names or default backgrounds, Preksha needs to select the appropriate background for scene generation. A single background image may have multiple attributes and at the same time, a single background object name may be tagged with multiple background images. Let us consider the following case: here, sentences inside a text have a reference of multiple words – शाम evening, लाल आसमान red sky, बादल cloud, सूरज sun, and नदी river, which all contribute in selecting the background. The sample annotated background images with these lexicon items in Preksha-R are shown in Figure 9.

The background information is mapped with OVF annotated objects in offline repository Preksha-R. Table 3 shows the sample entries for annotated object names mentioned in Figure 9. Here, "शाम" has 3, "आसमान" has 4, "बादल" has 4, "सूरज" has 2, and "नदी" has two references mapped with background images.

Table 3: Lexicon and object mapping in resource repository.

Annotated tag name	Attribute	File name
शाम, सड़क, पुल, सड़क-पोल, लाइट-बल्ब	Dark, light	Road
शाम, आसमान, बादल	Dark, red	dark-sky
आसमान, बादल, मैदान, जमीन	White, blue, green	whiteCloud
आसमान, बादल, मैदान, सूरज, धूप	Colorful, green	rivereve
नदी, पहाड़ी, पर्वत, किनारा, पत्थर	Blue, green	Lake
शाम, नदी, आसमान, बादल, सूरज	Dark, red	Ground

The most appropriate image named "dark-sky.jpg" with various references (शाम "evening," लाल आसमान "red sky," बादल "cloud," सूरज "sun," and नदी "river") of background information through multiple sentences of a text. Though it is not compulsory for a "chair" to have "room" as a background, Preksha system picks up a default background room (which is tagged with a lexical entry of "chair") in the absence of background information mentioned explicitly. The visual parameter of model "chair" helps in assigning the default background of object like – "study chair" is a more probable to be in "room" and "beach chair" at "sea beach" likewise "office chair," "lounge chair," "deckchair," and so on. Along with this, temporal information in input text also aids to enrich background preparation. It supports in augmenting "morning," "evening," or "night" time using the light effect in the scene. Behavioral information like "in strong wind" hints to have a background of the jungle, "In the sun" for "open ground," "traffic" shows "road," and "crowd" shows some "market" or "fair"; however, this assumption may not be true always.

In case of missing linguistic information or 3D artifacts of the objects from input text, a fall-back strategy is planned where user is provided with five options to proceed. A user interface prompts with various options: (1) continue without adding the missed object, (2) select a representative object from repository to proceed, (3) add basic temporary information of object to proceed, (4) add detailed linguistic and computer graphics information to be permanently in the repository, and (5) stop the

Preksha application without further executing. Based on the input received from end-user, the application decides how to proceed further.

7.1 Knowledge representation

The Preksha system is designed to produce a comprehensible and standard representation of this extracted knowledge, so that it can transform complete adequate information. The extracted knowledge from the parsed structure is captured and further represented in an XML file for scene planning. This KER is a self-sufficient structured data; therefore, it can also be used as input for any other scene generation engines, and not only for Preksha system.

7.2 Preparing scene knowledge

The scene knowledge is a collative information of (a) basic meta information of document – "MInfo," (b) environmental information related to the scene – "EnvInfo," and (c) entity information – "EntInfo" of all the 3D model objects in OAR form. Here, "p" is the number of objects to be rendered in the scene. Here, it can be expressed as scene knowledge = MInfo, EnvInfo, (EntInfo1, EntInfo2, ..., EntInfop). The extracted information from KEE engine is processed in class data structures. The instances of corresponding class objects are marshaled in XML format. The XML schema is explained further.

The tag <Scene> covers the entire information of the document with the knowledge of the instance of a virtual environment. After general basic meta information 'MInfo' about the XML document, the tag <Scene> has two subtags, namely, <EnvInfo> and <SceneInfo> (refer Figure 10(a)). Figure 10(b) describes the contents of <EnvInfo> and <SceneInfo>tag. The <EnvInfo> tag carries the environment information about the scene being rendered. It is a part of input information which is not very easily visualizable but necessary for adequate knowledge transfer. This is mostly the adverbial and temporal data extracted from the input text. The tag <EntInfo> arranges the knowledge about all the 3D Model objects along with all their required information to synthesize a plausible scene. The tag <EntInfo> organizes the data further in subtags like <Background> and <Objects> information. Figure 11(a) presents a sample for <Background> tag. The tag <Filename> holds the file name of background image. It is extracted using tagged lexicon in the linguistic repository with OVFs. Apart from the handling of a static background, which is explicitly mentioned in the input text, Preksha system takes care of additional processing to extract background information of scene.

For tag <Objects>, the word "object" refers to a physical entity and a 3D model in a scene. Here, one instance of <EntInfo> is a representation of one single object entity in the scene along with its attributes and relations. Therefore, each <Objects> tag is

```
<Scene>
  <Creator>"..."</Creator>
  <Generator>"..."</Generator>
  <Time>"..."<Time>
  <InputLang > "..."</InputLang>
  <InputText> "..."</InputText>
  <MetaInfo> "..."</MetaInfo>
  ...
  <MetaInfo> "..." </MetaInfo>
  <ScenetInfo>
    <BackGround> "..."
    </BackGround>
    <Objects> "..." </Objects>
    ...
    <Objects> "..." </Objects>
  </SceneInfo>
</Scene>
```

(a). XML Scheme

```
<Scene sceneType="Outdoor" numberOfSentences="2">
    <Creator>  Priyanka Jain  </Creator>
    <Generator> Preksha 1.1 </Generator>
    <Time> Sunday 8 April, 2021, 10:00 AM     <Time>
    <InputLang > Hindi </InputLang>
    <InputText> एक लकडी की कुर्सी और गोल टेबल धूप में है.
      गोल टेबल पर बैगनी तौलिया के ऊपर हरा फूलदान
      रखा है और उसके पास एक फूल सुबह से पडा है।
     </InputText>
 <EnvInfo> सुबह से </EnvInfo>
 <EntInfo>
     <BackGround> "..." </BackGround>
         <Objects> "..." </Objects>
         ...
         <Objects> "..." </Objects>
 </EntInfo>
 </Scene>
```

(b). EnvInfo and MInfo

Figure 10: (a) XML schema, and (b) EnvInfo and MetaInfo.

assigned with a unique-ID, which is an index of a sentence and token number into input text. The tag <Filename> is the 3D model object file name corresponding to noun mapped in Preksha-R. The tag <PropertyDefault> is an annotated feature using OVFs in Preksha-R. The <Objects> tag also has subtag <Relation> to represent relation name with its associated objects or parents. The tag <ParentID> is the index of the parent object in the input text. The Tag <ParentObjID> is the ID of the parent object, which helps in retrieving the name of the file to be associated. The <Position> tag is used to define its coordinates for object placement in virtual space. An object information tag for "flowerpot" is presented in Figure 11(b). For KER, sequencing of object placement

```
<BackGround id = "1_2" object="Obj-000Room"
    type ="Background">
<Filnename>room08</Filename>
<PropoertyDefaut>bg=1/bg, type=0/type, V=1, C=1,
    cp=Prop-CLR101/cp</PropoertyDefaut>
<Property />
<Relation />
<ParentID />
<ParentObjID />
<Position />
<BackGround>
```

(a). Background Tag for 'Room'

```
<Object id = "4_24" object="Obj-0060Pot"
    type ="Supported-by">
<Filnename>Pot_golden</Filename>
<PropoertyDefaut>1, 1, 0, (1.1, 1.2, 1.7)
    Prop-Clr, 0, 1, 0, 1</PropoertyDefaut>
<Property />
<Relation>RP_005_UP</Relation>
<ParentID>4_22</ParentID>
<ParentObjID>Obj_001_Table</ParentObjID>
<Position />
<Object>
```

(b). Object Tag for 'Pot'

Figure 11: Background tag for room.

is analyzed as per their positioning dependency value and roles mentioned in OVFs as shown in Table 1.

8 Marshaling the XML

The XML is the self-proclaimed universal format for structured documents and data. The XML allows users to define custom "tags" and "elements for the tags" which allows it to represent any content. In Preksha, the XML is used as the intermediary medium for communication between the separate modules, as well as the structure of the knowledge base. The Preksha's KER is done using JAXB API, where mapping the Java class to XML is done by creating a suitable serialized JAXB-compatible object. The marshaling in JAXB is the process of transforming a Java object into an XML file. The unmarshaling transfers an XML file into a Java object. The JAXB requires @ XmlRootElement annotation on the top most class which we are going to marshal or unmarshal. The Java object needs to be an instance of Java class. The JAXBContext represents a collection of JAXB-enabled classes, by which marshaller and unmarshaler objects are obtained for actual serialization. The benefit of using JAXB is that it validates XML schema automatically during the transformation process itself. This technique guarantees that all element tags match correctly, therefore, ensuring the well-formed structure of the XML parse tree.

9 Result

For scene synthesis process, the knowledge is extracted and represented by role matching and behavior assignment to the identified objects from the input text. The graphical constraints that represent the orientation, size, color, and texture of objects in the scene are obtained from the OVF set. The placement algorithm is done based on dimensions and relative positions of the objects using PDV. The KEE and KER are done using JAXB API which has control to bind with Java and additional validation capabilities for XML. In this way, the Preksha knowledge engine output is obtained in a standard form which is supported for all W3C XML Schema features. Figure 12 represents an extracted knowledge in XML form clause TC_b. Figure 13 is a synthesized scene corresponding to KER for clause TC_b as shown in Figure 12. This is done by scene planning and scene synthesis process. The scene planning transforms the knowledge output XML into WRL file format of VRML. The scene synthesis process uses Java3D API for SceneGraph processing in computer graphics.

The presented design has adaptability because it synthesizes a scene with rich environment, that is, a room as a background for mentioned objects in TC_b. Default background information is tagged with semantic feature-annotated objects in PrekshaR. However, it is not compulsory for a particular object to have a specific only one background. But in the absence of background information explicitly mentioned in the input text, Preksha system picked up a default appropriate background that is tagged with one of the lexical entries for noun object. The final scene is synthesized with background preparation, object selection, spatial relation, merging, positioning, camera view adjustment, and scene rendering, by using the knowledge engine output.

The scene synthesis corresponding to the full sample input text is presented in Figure 13. This is to mention here that the background of a scene is changed from "room" (Figure 13 of TC_b) to "sunlight" (Figure 14 of full text) in final scene synthesis. A complete spatial reasoning is handled at document level in the design of a knowledge engine. For example, in TC_c "हरे फूलदान के पास एक फूल सुबह से पड़ा हुआ है," the relative location of "हरे फूलदान" (green flowerpot) and "फूल" (flower) is not explicitly mentioned in final scene synthesis. However, it is positioned on the "तौलिया" (towel) that is above the "टेबल" (table) taking reference of TC_a and TC_b.

The Preksha KER is factual, as all the mentioned information presented in an input text is converted into final scene synthesis. For adequacy of knowledge, the temporal information present in the input text is also transformed into the rendered scene. This information is shown in the form of written text as a subtitle on the background of the scene. This is to note here that the generated scene is rich with ambiance grounding. The Preksha system has generated appropriate scenery, even in the absence of explicit mentioning of "chair with red back-support" and "field of grass with sunlight." The rendered 3D object models in the scene are picked up from Preksha repository as per their matching filename in the XML KER. This phenomenon

```xml
<?xml version="1.0" encoding="UTF-8" standalone="yes"?>
<Scene userName="" dateTime="" numberOfSentences="0">
    <InputText> गोल टेबल पर बैंगनी तौलिया के ऊपर हरा फूल-दान रखा है </InputText>
    <SceneInfo>
        <BackGround id="0_0" object="Default" type="BackGround">
            <FileName>room1.jpg</FileName>
            <PropertyDefault></PropertyDefault>
        </BackGround>
        <Objects id="4_33" object="Obj-02100" type="Object">
            <FileName>green-flower-bowl</FileName>
            <PropertyDefault>V=1, C=0/C, (x=1.7y=1.7z=1.7),</PropertyDefault>
            <Property>Prop-CLR-0 0.5 0_GREEN</Property>
            <Relation>RP-0002</Relation>
            <ParentID>4_30</ParentID>
            <ParentObjID>Obj-02129</ParentObjID>
        </Objects>
        <Objects id="4_30" object="Obj-02129" type="Container">
            <FileName>towel</FileName>
            <PropertyDefault>V=1, C=1/C, (x=8.7y=5.1z=8)</PropertyDefault>
            <Property>Prop-CLR-1 0 0_PURPLE</Property>
            <Relation>RP-0002</Relation>
            <ParentID>4_27</ParentID>
            <ParentObjID>Obj-0012TABLE_R</ParentObjID>
        </Objects>
        <Objects id="4_27" object="Obj-0012TABLE_R" type="Container">
            <FileName>Table_round</FileName>
            <PropertyDefault>V=1, C=1/C, (x=6y=4z=8), tx=Prop-TXT-WODDEN/tx,
            sp=Prop-0002ROUND/sp, BGD=room1/BGD, </PropertyDefault>
            <Property>Prop-0002ROUND</Property>
         </Objects>
        <Objects id="0_0" object="Default" type="BackGround">
            <FileName>room1.jpg</FileName>
            <PropertyDefault></PropertyDefault>
        </Objects>
    </SceneInfo>
</Scene>
```


Figure 12: Knowledge representations for TC_b.

is similar to the human mental imagery process, where the visualization of each individual is unique as per his/her prior observations and learning.

This data design with ID mapping helps in deriving the structured knowledge of objects in a language-independent form; hence, it is extensible to other languages. Also, this knowledge engine is independent of the parsing complexity in the Hindi language as it works on the output of the NLP engine. The final knowledge is formulated into a standard XML structure which can be used by any other engines and technologies. It is extensile for the further scope of research enhancement by only adding up more objects, attributes, and relations into the resource repository. The adequacy and sufficiency are the testing parameters for a knowledge engine. It has generality, that is, it represents and reasons over a broad range of spatial knowledge within its domain. It is competent so it produces correct and efficient results. It has

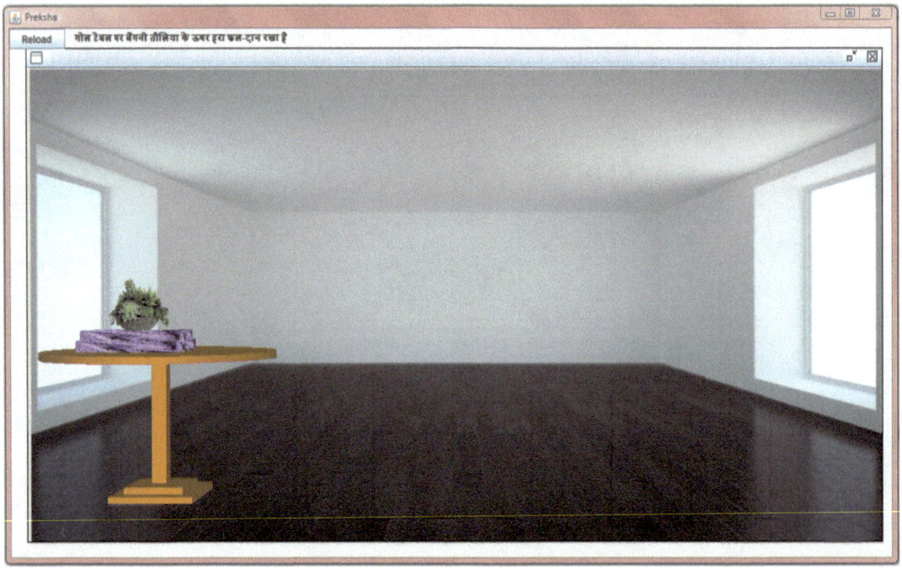

Figure 13: Scene generated based on clause TC_b.

Figure 14: A final scene generated for sample input text.

semantic clarity by having a clear and well-defined semantics. It can be used for other purposes; therefore, it has transformability.

10 Conclusion

We have designed and developed Preksha system for converting a given text in Hindi language to a scene. Preksha consists of three main engines: a language engine, a knowledge engine, and a scene engine. In this chapter, we have discussed the details of our knowledge engine. The knowledge engine receives its input from the language engine and generates output for the scene engine. This chapter presented details of the knowledge engine and gave design of a knowledge base that is used by the scene engine. We have discussed some of the challenges in an ATV system. Further, we have highlighted complexities of the Hindi language when designing an ATV system for it. We observed that the problem of linguistic ambiguities and mental imaginaries multiplies the complexities of an ATV system. A few illustrative examples are presented to demonstrate the processing in the knowledge engine.

We have defined OVFs to support the role assignment and behavior matching of the scene constituents. The requirement and importance of DoV are explored. The KEE at the text, clause, and local word group level is given. The KER process is explained for structuring the information about the physical objects mentioned in the input text, along with their associated attributes and relations.

The focus of this chapter is to visualize text which references physical objects visible to humans. The text containing abstract information, theories, and kinds of literature is not considered here. Although Preksha input text is at a sentence and paragraph level, with limited discourse information, it is modular and extensible for more complex input text. As the representation of knowledge engine is in the XML form, it is independent of natural languages and technology constraints. Preksha can be extended for information extraction, question answering, and natural language generation (storytelling) systems.

References

[1] Adorni, G., Di Manzo, M. & Giunchiglia, F.: Natural language driven image generation. In Proceedings of the 10th International Conference on Computational Linguistics and 22nd annual meeting on Association for Computational Linguistics. pp. 495–500. Association for Computational Linguistics. Stanford, California, 1984.

[2] Bharati, A. & Sangal, R.: Parsing free word order languages in the Paninian framework. In Proceedings of the 31st annual meeting on Association for Computational Linguistics. pp. 105–111. Association for Computational Linguistics. Columbus, Ohio. 1993.

[3] Bhat, A. R.: Exploiting Linguistic Knowledge to Address Representation and Sparsity Issues in Dependency Parsing of Indian Languages. Ph.D in. Computational Linguistics, IIIT-Hyderabad, India. 2017.
[4] Chang, A. & Savva, M. M. C. D.: Learning spatial knowledge for text to 3D scene generation. In Proceedings of the 2014 Conference on Empirical Methods in Natural Language Processing (EMNLP). pp. 2028–2038. Association for Computational Linguistics. Doha, Qatar. 2014.
[5] Clay, S. R. & Wilhelms, J. (1996). Put: Language-based interactive manipulation of objects. IEEE Computer Graphics and Applications, 16(2): 31–39.
[6] Coyne, B. & Sproat, R.: WordsEye: An automatic text-to-scene conversion system. In Proceedings of the 28th annual conference on Computer graphics and interactive techniques, SIGGRAPH 01, Computer Graphics Proceedings. pp. 487–496. New York, USA. 2001.
[7] Dupuy, S., Egges, A., Legendre, V. & Nugues, P.: Generating a 3d simulation of a car accident from a written description in natural language: The carsim system. In Proceedings of ACL Workshop on Temporal and Spatial Information Processing TASIP '01, Volume 13, Article No. 1, pp. 1–8. Association for Computational Linguistics. PA, USA. 2001.
[8] Earley, J.: An Efficient Context-Free Parsing Algorithm. In Magazine Communications of the ACM, Volume 13 Issue 2, pp. 94–102. 1970.
[9] Daniel, G. & Daniel, J. (2002). Automatic Labeling of Semantic Roles. In Computational Linguistics, 28(3), 245–288. Issue.
[10] Hassani, K. & Lee, W. S. (2016). Visualizing Natural Language Descriptions: A Survey. In ACM Computing Surveys (CSUR) Surveys Homepage archive, 49(1), Issue Article No. 17.
[11] Hermann, H. (2006). Knowledge Representation and the Semantics of Natural Language, Springer-Verlag Berlin Heidelberg, ISBN10 3-540-24461-1. 647.
[12] Jain, P. & Pawar, P.: From 'Pre-Position' to 'Post-position, In International Journal of Modern Computer Science (IJMCS) ISSN: 2320-7868 (Online), Volume 4, Issue Oct, pp. 66–71. 2016. (http://www.ijmcs.info/current_issue/IJMCS161034.pdf)
[13] Jain, P., Bhavsar, R. P., Kumar, A., Pawar, B. P. & Darbari, H.: Tree Adjoining Grammar based Parser for a Hindi text-to-scene conversion system. In 4th International Conference for Convergence in Technology (I2CT). 2018. (https://ieeexplore.ieee.org/document/8529491)
[14] Jain, P., Bhavsar, R. P., Lele, A., Kumar, A., Pawar, B. P. & Darbari, H.: Knowledge acquisition for automatic text visualization. In National Conference on Advances in Computing (NCAC2017). 2017 (http://ieeexplore.ieee.org/stamp/stamp.jsp?tp=&arnumber=7375651)
[15] Jain, P., Bhavsar, R. P., Pawar, B. P. & Darbari, H. (2018). Empirical Evaluation for Hindi text-to-scene generation system. International Journal of Creative Research Thought (IJCRT), I'2018 IJCRT | ISSN: 2320-2882 | Volume 6(1), 662–667. Issue http://www.ijcrt.org/IJCRT1872110.pdf
[16] Jain, P., Bhavsar, R. P., Pawar, B. P. & Darbari, H. (1-10. 2018). VRML for automatic generation of 3D Scene. In International Journal of Computer Application (2250-1797), 2(8), https://rspublication.com/ijca/2018/april18/1.pdf.
[17] Jain, P., Darbari, H. & Bhavsar, V. C.: Vishit: A Visualizer for Hindi Text. In Fourth International Conference on Communication Systems and Network Technologies (CSNT), pp. 886–890. IEEE Xplore. 2014. (http://ieeexplore.ieee.org/document/6821527/)
[18] Jain, P., Darbari, H. & Bhavsar, V. C.: Cognitive support by Language Visualization: A case study with Hindi Language. In 2nd International Conference for Convergence in Technology (I2CT), pp. 110–115. IEEE Xplore. 2017. (http://ieeexplore.ieee.org/abstract/document/8226104/)
[19] Jain, P., Darbari, H. & Bhavsar, V. C.: Spatial Intelligence from Hindi Language Text for Scene Generation. In 2nd International Conference for Convergence in Technology (I2CT), pp. 132–138. IEEE Xplore. 2017. (http://ieeexplore.ieee.org/abstract/document/8226108/)

[20] Jain, P., Darbari, H. & Bhavsar, V. C.: Text Visualization as an Aid to Language Learning Disability. In ELELTECH 2013 National Conference on e-Learning and e-Learning Technologies, India. pp. 88. 2013.
[21] Jain, P., Pawar, P., Koriya, G., Lele, A., Kumar, A. & Darbari, H.: Knowledge acquisition for Language description from Scene understanding. IEEE International Conference on Computer, Communication and Control (IC4-2015) Conference. IEEE Xplore. 2015.
[22] Joshi, A. K. (1987). An Introduction to Tree Adjoining Grammars, Manaster-Ramer, A. editor, Mathematics of Language, John Benjamins, Amsterdam.
[23] Lu, R. & Zhang, S. (2002). Automatic Generation of Computer Animation Using AI for Movie Animation, LNCS 2160 NY, Springer-Verlag Berlin Heidelberg, 1–374.
[24] Ma, M. (2006). Automatic conversion of natural language to 3D animation. Ph.D. thesis, Derry, Ireland, University of Ulster, 1–250.
[25] Pustejovsky, J. & Boguraev, B. (1993). Lexical knowledge representation and natural language processing. Artificial Intelligence, 63(12), Issues 193–223.
[26] Tappan, D.: Knowledge-Based Spatial Reasoning for Scene Generation from Text Descriptions. In Proceedings of the Twenty-Third AAAI Conference on Artificial Intelligence. Association for the Advancement of Artificial Intelligence (www.aaai.org). pp. 1888–1989. 2008.
[27] Shankar, V. (1987). A Study of Tree Adjoining Grammars. PhD thesis, Department of Computer and Information Science, Philadelphia, PA, University of Pennsylvania.
[28] Winograd, T.: Understanding Natural Language. Academic Press, Inc. Orlando, FL, USA. pp. 1–191. 1972. Also published in Cognitive Psychology, Volume 3, Issue 1, pp. 1–191, 1972. ISBN: 0127597506
[29] XML – https://www.w3.org/XML/ (Accessed on 10 Feb, 2018)
[30] Zeng, X., Mehdi, Q. H. & Gough, N. E.: Shape of the Story: Story Visualization Techniques, In Proceedings of the Seventh International Conference on Information Visualization (IV03). London, United Kingdom, IEEE. pp. 144. 2003
[31] JAVA3D: http://www.java3d.org/ (Accessed on 10 Feb, 2018)
[32] Wang, Y.: The OAR model for Knowledge representation, Proceedings of Canadian Conference on Electrical and Computer Engineering, IEEE Xplore. pp. 1727–1730. 2006.
[33] JAXB - https://docs.oracle.com/javase/tutorial/jaxb/intro/ (Accessed on 10 Feb, 2018)
[34] Yosuke, T., Hideo, S. & Masahiro, T.: Story Driven Animation System (SDAS), Proceeding, CHI '87 Proceedings of the SIGCHI/GI conference on Human factors in computing systems and graphics interface, Toronto, Ontario, Canada, pp. 149–153. 1987
[35] Scene Graph Basics, Java 3D API Specification. https://docs.oracle.com/cd/E17802_01/j2se/javase/technologies/desktop/java3d/forDevelopers/j3dguide/SceneGraphOverview.doc.html
[36] Spika, C., Schwarz, K., Dammertz, H. & Lensch, H. P. A. (2011). AVDT – Automatic Visualization of Descriptive Texts, Vision, Modeling, and Visualization, Eisert, P., Polthier, K. & Hornegger, J. Eds., The Eurographics Association, 129–136.
[37] Dakwale, P., Mujadia, V. & Sharma, D.: A Hybrid Approach for Anaphora Resolution in Hindi, International Joint Conference on Natural Language Processing, pp. 977–981, Nagoya, Japan, pp. 14–18. 2013.
[38] Gupta, C., Jain, A. & Joshi, N. (2018). Fuzzy logic in natural language processing–a closer view. Procedia computer science, Jan 1(132), 1375–1384.
[39] Gupta, C., Jain, A. & Joshi, N. (2019). DE-ForABSA: a novel approach to forecast automobiles sales using aspect based sentiment analysis and differential evolution. International Journal of Information Retrieval Research (IJIRR), Jan 1 9(1), 33–49.
[40] Gupta, C., Jain, A., Joshi, N. & Novel, A. Approach to feature hierarchy in Aspect Based Sentiment Analysis using OWA operator. In Proceedings of 2nd International Conference on Communication, Computing and Networking 2019 (pp. 661–667). Springer, Singapore.

[41] Bertin, J. (1983). The Semiology of Graphics, Berg, W. J. trans University of Wisconsin Press, Madison, xi, ESRI. ISBN 978-1-58948-261-6 415.

[42] Ashima, A., Kaur, S. B. & Mohana, R. C. (2016). Anaphora Resolution in Hindi: A Hybrid Approach, Advances in Intelligent Systems and Computing, Springer International Publishing AG.

[43] Coyne, B., Klapheke, A., Rouhizadeh, M., Sproat, R. & Bauer, D.: Annotation Tools and Knowledge Representation for a Text-To-Scene System in Proceedings of COLING 2012, pp. 679–694, 2012.

[44] Logan, G. D. & Sadler, D. D. (1996). A computational analysis of the apprehension of spatial relations in Language and space, Bloom, P., Peterson, M. A., Nadel, L. & Garrett, M. F. Eds, Cambridge, MA, US, The MIT Press, 493–529.

[45] Coyne, B., Sproat, R. & Hirschberg, J.: Spatial relations in text-to-scene conversion, in Workshop at Spatial Cognition: Computational Models of Spatial Language Interpretation, Mt. Hood, OR, USA, pp. 916. 2010.

[46] Coyne, B.: VigNet: grounding language in graphics using frame semantics, Proceedings of the ACL 2011 Workshop on Relational Models of Semantics, pp. 28–36. 2011.

[47] Miller, G. (1998). WordNet: An Electronic Lexical Database, Cambridge, Mass, A Bradford Book.

[48] Jain, P., Shaikh, K., Bhavsar, R. P., Kumar, A., Pawar, B. P., Darbari, H. & Bhavsar, V. C.: Cascaded finite-state chunk parsing for Hindi language. In: Proceedings of international conference on communication and information processing (ICCIP-2019) and in Elsevier-SSRN 2019 (https://papers.ssrn.com/sol3/papers.cfm?abstract_id=3422328)

[49] Jain, P., Jain, N. K. & Darbari, H., प्रेक्षा: भाषा प्रेक्षण (विज़ुअलाइज़ेशन) द्वारा संज्ञानात्मक सहयोग, 'VIGYAN PRAKASH: Research Journal of Science & Technology,' UGC-CARE Journal ISSN: 1549-523-X, Vol: 17. 2019 (http://vigyanprakash.in/article.php?id=17.1.4)

[50] Jain, L. & Agrawal, P. (2015). "Text Independent Root Word Identification in Hindi language using Natural Language Processing". International Journal of Advanced Intelligence Paradigm (IJAIP), 7(3/4), 240–249. Inderscience Publications.

[51] Jain, L. & Agrawal, P., "Sheershak: An Automatic Title Generation Tool for Hindi Short Stories", International Conference on Advances in Computing, Communication Control and Networking (ICACCCN'18), pp 579–584, IEEEXplore, 2018.

[52] Jain, P., Bhavsar, R., Shaikh, K., Kumar, A., Pawar, B. V., Darbari, H. & Bhavsar, V. C.: Virtual Reality: An aid as Cognitive Learning Environment. A case study of Hindi Language, in Special Issue of XR (VR, AR, MR) and Immersive Learning Environments, in Springer Nature Journal "Virtual Reality" ISSN: 1359-4338 (Print) 1434–9957. 2020 (https://link.springer.com/article/10.1007/s10055-020-00426-w)

[53] Jain, P., Bhavsar, R., Shaikh, K., Kumar, A., Pawar, B. V., Darbari, H. & Bhavsar, V. C.: Evaluation of Automatic Text Visualization Systems: A Case Study. in "Springer series of Advances in Intelligent Systems and Computing" and 5th International Conference on Advanced Machine Learning Technologies and Applications (AMLTA-2020)., vol 1141. Springer, Singapore. 2021 (https://doi.org/10.1007/978-981-15-3383-9_3)

[54] Jain, P.: Preksha: A Hindi Text Visualizer, Ph D. Thesis, NMU, Maharashtra India. March, 2020.

[55] Jain, L. & Agrawal, P.; Sheershak: An Automatic Title Generation Tool for Hindi Short Stories, 2018 International Conference on Advances in Computing, Communication Control and Networking, 312–321, 2019.

Index

https://doi.org/10.1515/9783110713633-013